CATASTROPHE
ETHICS

CATASTROPHE ETHICS

How to Choose Well
in a World of Tough Choices

TRAVIS RIEDER

DUTTON

DUTTON

An imprint of Penguin Random House LLC
penguinrandomhouse.com

LIBRARY OF CONGRESS CATALOGING-IN-PUBLICATION DATA

Names: Rieder, Travis N., author.
Title: Catastrophe ethics: how to choose well in
a world of tough choices / Travis Rieder.
Description: New York : Dutton D, [2024] |
Includes bibliographical references and index.
Identifiers: LCCN 2023022647 (print) | LCCN 2023022648 (ebook) |
ISBN 9780593471975 (hardcover) | ISBN 9780593471999 (ebook)
Subjects: LCSH: Decision making—Moral and ethical aspects. |
Moral conditions. | Ethics.
Classification: LCC BJ1419 .R54 2024 (print) | LCC BJ1419 (ebook) |
DDC 179.7—dc23/eng/20231205
LC record available at https://lccn.loc.gov/2023022647
LC ebook record available at https://lccn.loc.gov/2023022648

Printed in the United States of America

1st Printing

Book design by Nancy Resnick

To my family.
Thank you for providing me
with a little corner of the world
free from catastrophe and full of joy.

Contents

PART III
SOLVING THE PUZZLE

PART IV
CATASTROPHE ETHICS

CATASTROPHE
ETHICS

May I be the tiniest nail in the house of the universe, tiny but useful.

—Mary Oliver, *Upstream: Selected Essays*

It's Hard to Be Good.
And It's Getting Harder.

People who try hard to do the right thing always
seem mad.

—Stephen King, *The Stand*

Modern life is morally exhausting. And confusing. Everything we do seems to matter. But simultaneously: nothing we do seems to matter. My friend, an outspoken environmentalist, recently posted to social media a picture of herself on a beautiful beach, celebrating a moment of quiet with nature. And predictably—the internet being what it is—within a few moments of posting it, one of the first comments came in: "How much carbon did you emit to go on your vacation?" The implication, of course, being that she's a hypocrite, preaching environmentalism for thee, but not for me. And despite the comment seeming like a childish jab, she—like most of us—cares about justifying her actions, and so she responded, citing all of the ways that she minimizes her carbon footprint and arguing that never getting to enjoy her life seems like an unreasonable standard.

1

This sort of debate plays out in my head, with me playing both sides, regularly—many times a day, if I let it. This morning at breakfast, I poured almond milk on my cereal, which is the result of a judgment I made years ago when I decided that cow milk is too environmentally expensive to justify. In general, animal-based foods have a higher carbon footprint than their plant-based counterparts, and so I have, to varying degrees over the years, reduced or eliminated them from my diet. But while working on a recent food ethics project, I learned that almond milk might not be a great substitute. Although it does have a lower carbon footprint, almond trees require huge amounts of water—like, twelve liters of water to produce a single nut—and more than 80 percent of the world's almonds are grown in California, which suffers from severe drought. Switching from cow milk to almond milk thus traded a high carbon footprint for high water use. So I sat down to research soy milk and oat milk to see if I could learn which was the least harmful product, ultimately deciding that none of them are perfect but that all plant-based alternatives are superior to cow milk. To some degree, it seems to matter which milk I use, since it contributes to huge problems like climate change and water scarcity. That's why companies advertise their green bona fides when they have them. But then again, we're just talking about a little milk on my cereal. Should I actually stress about this? Doing so starts to feel a bit precious, a bit obsessive, a bit righteously focused on my own purity.

I also drove to the gym today, which reflects many ethically relevant decisions I've made about my life. I live in the suburbs, which means I have to own a car and drive most places I want to go. Making this choice supports a lifestyle popular in America that is very bad for the environment—one of spread-out individualism, in which so many of us live in big houses, with big, mono-crop lawns, driving our private cars that carry only us as single passengers to do every minor task. My drive to the gym or my forty-five-minute commute to campus is a reminder that I'm part of a radically unsustainable cultural choice.

I do, however, try to minimize the effect of this lifestyle by owning an electric vehicle and driving it rarely. I work from home when I can, and most trips are less than ten or fifteen miles. So I've responded to the sense of being implicated in a problematic structure by trying to minimize my participation in it. But I know it's not a perfect response, and so I feel moderately guilty about my suburban home and my private car.

Participating in our broader consumer culture also seems to carry moral weight. I just received a book I ordered in the mail, and I have complicated feelings about that. Online shopping is phenomenally convenient, but it comes with costs. I would have preferred to support one of my favorite local bookstores, even though that would bring with it some extra driving, but I didn't have time to make the trip (from here in the suburbs of course) before I needed the book. And while that means I didn't need to make a trip, the delivery person did, coming all the way to my home to deliver a box with just a single book. I have no idea how to feel, morally, about all the various components of this small consumer decision.

Even our entertainment decisions are not left untouched by such moralizing. In recent years, there have been extensive debates about the appropriateness of "deplatforming," or boycotting problematic artists. In the wake of the #MeToo movement, it became clear that some previously beloved actors, directors, comedians, and other public figures were deeply problematic people. Some of them not only sexually harassed or assaulted women, but did so for years with impunity—becoming unbelievably wealthy and powerful the whole time. When we learn this about someone, should we stop consuming their products? It feels like we should, though it's a bit hard to articulate why. If I stream a stand-up comedy special through Netflix, it's not likely that an artist (or their financial backers) would notice that I viewed their product, nor would my refusal to watch have any measurable impact on them. And some people argue that we should separate a person from their art. After all, if someone is a misogynist or racist, it's not clear that this makes their art less beautiful, funny, or entertaining. But watching does seem to support them in some way, even if they wouldn't notice, and that support feels morally problematic.

A similar point came up in late 2022 when Qatar hosted the FIFA World Cup. The matches making up the World Cup are some of the most watched events in the world, with an audience numbering in the billions. Even those of us who don't closely follow soccer (or football, as our international friends would have it) are drawn into the high-drama matches. But the 2022 matches were shadowed by significant controversy. When Qatar won the bid to host the Cup, it had nowhere near the infrastructure necessary for such a massive event. As a very small country—approximately 80 percent of the size of Connecticut—Qatar needed to essentially build an entirely new city in which to hold the tournament, including stadiums, hotels, restaurants, and transportation. But with a population of only 300,000 Qataris, the burden of this work fell on the millions of migrant workers brought in from Asia. In recent years, it has become clear that this system of employment was widely abused, with immigrants working long hours in the blazing heat and then living in stifling, unsanitary housing camps. A report by *The Guardian* concluded that in the decade after being awarded the World Cup hosting rights in 2010, more than 6,500 migrant workers from South Asia died in Qatar. Although FIFA and Qatar dispute this number, there are ample reports of atrocious living and working conditions, making it clear that, whatever the precise number of people who died from the attempt to rebuild a nation in just over a decade, it was a terrible burden that fell on a vulnerable population.

So should none of us have watched the World Cup? I certainly felt the pull in that direction. To watch was to support FIFA's and Qatar's decision to build a city in a desert with exploited migrant labor, which predictably caused terrible suffering. But at the same time: no one cares whether I watch or not. When billions of others are watching, my choice is insignificant. And the World Cup is a cultural phenomenon that carries a lot of meaning for many people. Should a Lionel Messi fan really be expected to give up watching his last dance in order to take some sort of righteous stand?

We could continue digging up cases. But the basic context that makes up the confusing ethics of today is the following: many of us feel an individual responsibility to address massive, collective problems, despite an inability to act in ways that have a meaningful impact on those problems. We, as a global society, must address climate change. Doing so is an absolute moral requirement because it is already causing devastation, with the worst yet to come. As a result, I feel some responsibility not to contribute to climate change, and to be a force for positive change, even though, realistically, my contributions either to the problem or to a solution don't seem meaningful. The problem is too big, and my contribution too small, to make a difference. I'm torn between the pull of a kind of purity (that I should keep my hands clean by withdrawing from problematic activities) and a sense of nihilism (that it doesn't matter what I do, so I should get over myself and just live my life).

So what is each of us to do? How do we live a morally decent life when we can't even get our arms around the problems? That's the central question of this book.

The psychologist and philosopher Joshua Greene thinks the human brain is like a camera: it has an automatic (fast, easy to use, not very versatile) setting and a manual (slow, effortful, but versatile) setting. Similar to the psychologist and economist Daniel Kahneman, Greene thinks that this leads us to make different kinds of judgments in different situations, and that this extends to moral judgments as well. Just like we have fast intuitions about danger (snakes!) and utilize slow deliberation in other situations (how do you determine the volume of a sphere, again?), we have both types of judgments about moral problems. Our automatic moral camera is often in charge, making judgments quickly and unthinkingly, helping us to navigate the world without constantly slowing down and deliberating. And like with its non-moral counterpart, the fast judgments it makes are often correct. I don't typically need to deliberate about whether to lie or keep my promises, and I never have to wonder about whether I should cause random violence. Our auto setting is helpfully efficient.

To get reliable fast judgments, however, our automatic cameras must be calibrated. Greene tells us we can get such calibration through genetic transmission, cultural transmission, or effortful learning. Genetic transmission likely explains some of our deepest fast judgments (Snake! Danger!), and we learn many of them through cultural absorption as well (Look out! A gun!). But many of our automatic settings were calibrated through personal experience (A stove! Hot!).

If we never train our automatic setting at all, then we shouldn't expect it to produce reliable judgments. And this is the crucial point about moral intuitions. Sometimes we find ourselves in a novel situation—one that neither our ancestors nor our culture nor our personal experience has prepared us for—and yet we have a moral intuition about what to do. Should we trust this fast, automatic judgment? Greene says no, because there's no reason to think it's gotten the case right. When we find ourselves in unfamiliar circumstances, he says, it would be a cognitive miracle if our judgments were reliable. And since we should assume there are no cognitive miracles, we should distrust any immediate intuition in such a case and switch over to manual moral mode. If you had never seen a car before and yet you had an intuition about how to drive when you sat down in it, you should view that intuition with a healthy dose of skepticism. The ability to drive isn't passed down through our genes, so if you haven't absorbed it from culture and you haven't practiced, you're gonna have to kick it into manual mode in order to develop driving instincts. Once you've calibrated those automatic settings, then you may find that you can get in the car, pull away from work, and show up at home sometime later without putting any real cognitive effort into your commute. But you can't do that without driving lessons and lots of practice.

Many of the moral problems of today are unfamiliar. They are so different from the ones that humans have faced before that we shouldn't assume our moral cameras have been trained on a reliable set of data. Greene himself notes climate change as a paradigmatic instance of an unfamiliar moral problem, and it will serve as the motivating example for much of what is to come in our exploration. But it's not just climate change. Many of the problems we face now are radically different from those that existed for much of humanity's time on Earth.

Humans evolved in fairly small groups, where the effects of our actions could largely be witnessed or inferred. The moral rules that humans developed to regulate their actions thus made sense as they responded to the most salient ethical considerations of the world around them. These rules focused on concepts such as harm to individuals or the rights of individuals. Over the past two centuries, however, the scale and complexity of the world has grown almost unimaginably. In the year 1800, there were still fewer than 1 billion people alive, and they were spread around the world, without direct or easy access to one another. The incredible technological advances brought by the Industrial Revolution changed both of those features. The global population more than doubled to 2 billion by 1928, and societies became steadily more connected to and reliant on one another for trade and alliances. The next billion people were then added in just over thirty years, and the most recent 5 billion people were added during a single generation. Anyone alive today who was born before 1960 has seen the global population go from 3 billion to more than 8 billion, and has witnessed technological evolution that made the world, so much bigger in terms of people, feel so much smaller in terms of reachability.

It is in the context of this massive, global society of billions of people engaged in various loosely connected projects that the question of my individual contribution to catastrophic problems arises. Our ancestors' morality did not prepare us for climate change. And we as individuals or as a culture have not yet figured out how to respond to the moral demands of problems so grand in scope. The ethics of massive, structural problems—catastrophes that are largely insensitive to my individual choice—are unfamiliar, so they require that we switch our moral cameras over into manual mode.

Greene thinks that when we engage manual mode, we tend to use some of our ethical tools more than others, and that this serves as a kind of indirect argument in favor of his preferred moral philosophy. But the realization that we're in an unfamiliar ethical setting can have more radical implications than that. If the moral context of today is truly different from so much of human history, why think that our standard ethical theories will work at all? It may well be the case that adequately addressing genuinely novel moral questions will require articulating or discovering new moral principles, or focusing on different moral concepts than we are accustomed to. And, indeed, I think this is precisely what is needed in the face of problems like climate change. We tend to think about individual ethics in terms of concepts like harm, rights, and character, but an individual choice to make a relatively slight contribution to climate change does not by itself harm anyone, violate their rights, or explain why one's character is deficient.

The moral challenges of today are absolutely daunting, but they're also unfamiliar, which means we should be suspicious of any judgments that are a bit too quick and easy. We need to calibrate our fast moral judgments in an era of climate change and other global catastrophes, and this may require a broader set of ethical tools than we have typically employed.

I have divided the argument to come into four parts. The first part is my attempt to pay off the claim that (at least some of) the moral challenges of today are unfamiliar in the sense described by Greene. Climate change is the paradigm example of the puzzle of individual morality in the context of a massive collective challenge. What makes individual ethics difficult to think about in the case of catastrophic climate change makes ethics difficult to think about in many other contexts as well. "The Puzzle," as I put it, is everywhere.

One solution to the Puzzle is to insist that it's not that puzzling because our traditional moral reasoning can handle it in one way or another. In order to evaluate that claim, we need to know something about traditional moral reasoning, and so I explore that topic in Part II. The central question of that part of the book is, "How might classical moral philosophy solve the Puzzle?"

In Part III, we see that the answer is "Not well"—at least, that the tools most of us unthinkingly rely on when engaging in moral deliberation are not particularly helpful when it comes to reasoning about individual responsibility for large collective problems. We must expand our suite of ethical concepts if we want to satisfactorily solve the Puzzle. In Part IV, then, I detail what it might look like to change our thinking in the way I suggest, addressing both the mundane, everyday issues like whether to buy bottled water and the monumental moral question of whether to have children. Here, I offer a picture of how to live a morally decent life in the often-scary, always-complicated world we and our children now find ourselves in—an account of what I call Catastrophe Ethics.

PART I

THE PUZZLE

CHAPTER 1

The Climate Case

It is worse, much worse, than you think.

—David Wallace-Wells, *The Uninhabitable Earth*

I'd love to retire in Cyprus. It's as close to paradise as anywhere I've ever been, with long stretches of largely unspoiled beaches, warm Mediterranean water, mild winters, friendly people, and amazing food. My partner, Sadiye, is Turkish Cypriot (from the northern half of the island), and so we have a home there, full of people who love us. I love the Turkish language and the culture, and every time we go home, we wonder how long it will be until we can just stay there.

Unfortunately, I don't think we'll retire in Cyprus. Not year-round, anyway. It'll just be too hot.

Since it's an island sitting at the intersection of the Mediterra-nean, the Middle East, and Africa, you might expect it to be a warm climate. And that's how it used to be. But it's going from reasonably warm to unbearable in the summer months.

Among the hundreds of Cypriots I've met since my first visit in

2003, I've never encountered climate denialism there. There's not much room for doubt when the older generations have seen the environment go from hospitable, though warm, to downright hostile for two to three months of the year. Indeed, my in-laws regularly tell me about how much change they've seen in their lifetimes. Even when Sadiye was growing up, families largely didn't invest in expensive air-conditioning units (which use very expensive energy). She and her brother grew up sweating at night during the summer because it was at least possible to get away with forgoing A/C units in the children's rooms.

Not anymore. Every Cypriot home I go to now has A/C units in every bedroom, and most have the (prohibitively expensive) large units in their living space to get them through the worst of the summer months.

By the time Sadiye and I intend to retire—somewhere around the midcentury mark—I don't think we'll want to be in Cyprus for a good chunk of the summer. Of course, many Cypriot people will still live on the island, but it will be getting more uncomfortable to do so (and they will likely have increased the already prevalent tendency to simply go on holiday in August to escape the heat). My family and friends already have modified work schedules during the heat of the summer, which encourage people to stay indoors and find air conditioning during the hottest part of the afternoon. Construction and other manual labor slows down or stops. Those without air conditioning or who can't afford to run it constantly struggle to stay not only comfortable but healthy.

In fall 2021, it was announced that August of that year had been the hottest month ever recorded in Cyprus. High temperatures averaged 39.8 degrees Celsius (103.6 degrees Fahrenheit), with a single-day high temperature of 44.3 degrees Celsius

(111.74 degrees Fahrenheit). To see how quickly things are changing, consider that the average high temperature in August from 1981 to 2010 was 37 degrees Celsius (98.6 degrees Fahrenheit). And warming trends are only speeding up. Every Cypriot I have spoken to about climate change is deeply concerned about the future, as anyone my age or older has viscerally felt the dramatic change over their lifetimes.

When I was growing up and first learned of climate change, the wisdom of the day was that it occurred on a timescale too slow to notice. That was part of the challenge, in fact, that we needed to address a threat that one couldn't actually see over the course of a lifetime. But we've now come to realize that's simply not true, and we're already seeing the changes. People like me who spend our lives thinking about this issue will tell you: we make very practical decisions based on the future that climate change is ushering in.

Cypriot summer during my sunset years isn't the only time and place I'm avoiding. During the summer of 2020, it seemed like the entire West Coast of the United States was ablaze. Residents of California couldn't go outside due to air-quality warnings, and pictures of the San Francisco Bay Bridge were all over the internet, standing against a threatening, apocalyptic-looking orange sky. Fire season is spreading dramatically in California, both in duration and intensity. There is much that Sadiye and I love about the Bay Area: the weather; a strong biotech presence for her (she's an industry scientist); a thick network of universities for me. But my climate angst won't let us entertain the idea of moving there.

Unsurprisingly, another restriction on our future plans is that we avoid moving too close to the ocean. Living on the East Coast, moving somewhere along the eastern shoreline of the United States seems almost attainable. But the increasing number and intensity of tropical storms, along with sea-level rise, has me acutely aware of the ever-increasing costs (both financial and anxiety-related) of those beautiful water views. Iconic beach destinations like the Outer Banks in North Carolina are slowly being lost to the Atlantic Ocean, with beachfront homes now regularly falling into the rising waters and the only highway into the area constantly requiring protective adaptation (and a section of it being converted into a bridge, for a price tag of $155 million).

Miami-Dade County is perhaps in an even more dire circumstance, sitting on porous limestone on the low-lying southern tip of Florida. There, the phenomenon of "sunny-day flooding," which is flooding just from high tide, has become a standard nuisance, forcing the city to spend hundreds of millions of dollars in an attempt to prolong its life. According to the National Oceanic and Atmospheric Administration, the same story is playing out in coastal communities around the world. But adaptation efforts like seawalls, beach nourishment, building houses on stilts, and converting roads to bridges are Band-Aids. Low-lying coastal areas will eventually be overtaken by rising waters while being ever more routinely battered by ferocious storms.

If you're wondering whether it's a bummer to live with someone who spends their days thinking about climate change, Sadiye is probably too nice to say yes but too honest to say no.

Of course, much of this caution is based on predictions. Yes, Cyprus is already hotter, and yes, Miami has a "King Tide" season. But my worry that it will continue to worsen to a degree that should make most of us want to stay away is based on projections. So couldn't I be wrong? After all, we're hearing all the time about climate summits, new policies and pledges made by various countries, and green technologies that are supposed to revolution-ize our future.

Perhaps, then, I'm just being pessimistic. Humanity is going to figure this out, right? After all, the Netherlands figured out how to protect itself against the sea. When we realized that our collective actions were causing a hole to appear in the ozone, we moved quickly to adopt policies and change behaviors in a way that solved the problem. Surely we'll eventually pull off something similar when it comes to climate change. Right?

I want to tell you yes. I want to say that there's still hope that we will avoid the sort of future I fear. But while there's a technical sense in which we could, it's not realistic. Serious harms are already here, many more are coming, and we avoid talking about them at our peril.

For hundreds of thousands of years, the atmosphere was relatively stable in an important respect: the proportion of greenhouse gases (GHGs) such as carbon dioxide (CO_2), methane (CH_4), and ni-trous oxide (N_2O) remained within a certain range. That's impor-tant because these gases are very efficient at trapping heat, so when the concentration of GHGs in the atmosphere goes up, less heat is able to escape from Earth out into space and so the global average temperature increases. The atmosphere is like a blanket that keeps the Earth warm (this is good up to a point, since without it, the planet would be an icy, lifeless rock), and we can add insulation to that blanket by emitting more heat-trapping gases into it. Al-though all GHGs contribute to climate change, CO_2 gets the most attention, as it is the primary GHG emitted by humans; in the

United States, it constituted 79 percent of all GHG emissions in 2020.

Prior to the Industrial Revolution, the atmospheric concentration of CO_2 was around 280 parts per million (ppm), but the development of the world's postindustrial economies resulted in two changes in human behavior that had profound effects on the climate. First, it massively increased the amount of CO_2 we spew into the air (primarily through the burning of fossil fuels). And second, it reduced the planet's resources for dealing with that excess by engaging in practices like clear-cutting forests, destroying peatlands for agricultural or other purposes, and generally going about stripping the natural world and replacing it with a built one. These forests and peatlands that we've eliminated make up some of the world's most important "carbon sinks," which is one of the ways in which carbon is removed from the atmosphere.

So we've both filled the air with CO_2 and limited the planet's ability to regulate atmospheric concentrations.

The result is that the atmospheric concentration of CO_2 has been steadily climbing since the Industrial Revolution, with mean global temperatures right behind it. Although the science of climate change has been recognized for hundreds of years, the severity of the situation began to come to public attention in the late twentieth century, with the fundamental question being: How warm can we allow the planet to get before it causes serious, irreversible harm? And the important follow-up: How much CO_2 can we release into the atmosphere and stay below whatever that threshold is?

Although many took that first question to be a scientific one (involving predictions about the effects of warming on human populations), it's actually a moral one. We needed to know what the costs of warming were, and to whom, so that we could draw a line at some point and say that it would be wrong to let a certain amount of damage happen.

The answer that evolved and began to gain consensus at the end of the twentieth century was that we must keep warming below 2 degrees Celsius (global average). This number first appeared in a publication by Yale economist William Nordhaus in 1975, and slowly seeped into the scientific and political reasoning over the following two decades. In 1996, a limit of 2 degrees was cemented in policy aspirations by the European Council of Environment Ministers, which stated, "Global average temperatures should not exceed 2 degrees above pre-industrial level."

The answer to the second question took longer to answer. What must the human population do in order to prevent 2 degrees of warming? There is significant uncertainty here, but it looks like we can do that by keeping atmospheric CO_2 below 450ppm. Using this number, scientists were able to calculate an all-time, anthropogenic (human-caused) carbon budget. That is to say, there is a certain amount of carbon that we can emit if we want to stay below that 450ppm threshold, thus avoiding 2 degrees of warming. That number? One trillion tons of carbon, or approximately 3.67 trillion tons of CO_2.

A trillion metric tons. The number 1 with twelve zeros after it. An unbelievably large amount of something unimaginably small.

That's the budget humanity was given.

Despite the sheer magnitude of that number, we were quickly approaching it. Already in 2009 when the budget was first proposed, it was calculated that humanity had used more than half of it, and global emissions were still speeding up. Worse was the fact that 2 degrees' warming was not a goal that we should want to hit.

Scientists and politicians drew a line in the sand at 2 degrees warming, claiming that we should never *cross* that threshold; but that does not mean that it would be a good place to *land*. Two degrees' warming is not what we can allow without causing harm; it's what we can allow without causing massive harm at a magnitude that is essentially irreversible on a human time scale.

As a result, some scientists and activists began claiming that we had answered the moral question incorrectly: 2 degrees' warming is too lax; we should have aimed to prevent warming from exceeding 1.5 degrees Celsius. No one thought that limiting warming to 2 degrees would prevent harm, so we should look at who would be harmed if the planet were allowed to warm 2 degrees. And the answer, unsurprisingly, was that the world's poorest and most marginalized would be hit first and worst, with the globally privileged citizens largely able to protect themselves from such warming. So 2 degrees might be manageable by the most economically developed nations (though, to be clear, it would not be without real costs); but it would be devastating to low-lying Pacific Island nations and coastal regions, as well as some of the hottest areas, like the Middle East, North Africa, and parts of Asia and South America. As climate activists recalibrated their goals and began to bring more visibility to the risks of allowing 2 degrees of warming, the Paris Accords—which 196 nations adopted in December 2015—pledged to keep warming under 2 degrees while acknowledging that it would be better to keep warming under 1.5 degrees. If we want to make that new, more ambitious target, we must limit carbon not to 450ppm but 430ppm, which drastically cuts the time we have in which to respond.

How close to that limit are we now? In 2022, the Mauna Loa Observatory in Hawaii recorded a CO_2 concentration above

420ppm for the first time, and it's going up—fast. Over the last decade, the annual increase of CO_2 concentration has been more than 2ppm, which means that we're on pace to hit 430ppm CO_2 around the end of the decade, locking in 1.5 degrees Celsius warming.

So why all the optimistic talk by politicians, at climate summits, and on the news? Does a treaty like the Paris Accords indicate that we still have a chance at avoiding dangerous global warming?

Here's where the answer is technically yes, but not in a way that will make you feel any better.

The Intergovernmental Panel on Climate Change, or the IPCC, is the international body of the world's leading scientists which, every several years, puts out a new assessment report detailing the state of climate science, predictions about the future, and policy proposals that could change those predictions. In short, they tell us how bad things are and what we can do about it.

Over the course of 2021 and 2022, the IPCC put out its Sixth Assessment Report, which continued the trend of the previous five reports getting more alarming with each edition. The science is getting more mature, our predictions are getting more accurate, and the picture presented by that science and those predictions is getting scarier.

However: the IPCC tells us there are still ways to avoid 2 degrees Celsius global average temperature rise, and in principle, even a way to avoid crossing 1.5 degrees by too much. It's difficult, and it gets harder with every day we wait. But it's possible.

That sounds like good news. And it is! But it undersells just how difficult it will be to live up to those best-case scenarios. The fact remains that mitigating climate change requires massive political action, and the nations of the world are continuing to drag their feet. Despite the celebrations surrounding the Paris Accords and the recognition of the need to limit warming to 1.5 degrees, political action has lagged significantly. The most recent UN Emissions Gap Report shows that nations are not on track to meet their emissions reduction pledges, and that those pledges are themselves far too weak to avoid dangerous warming. While the global community needs to decrease emissions by 45 percent by 2030 compared to the 2010 level, current projections indicate that emissions are expected to *increase* by 10 percent in that time.

Yet another reason to worry is that the road map the IPCC gives us to limit warming requires some actions and policy choices that may be either politically controversial or even technically impossible. On the political front: one assumption that gives us a chance to avoid 2 degrees' warming is that many nations will aggressively scale up nuclear power. Why? Because nuclear power is clean energy from a GHG standpoint. Nuclear power plants are able to provide huge amounts of energy for a very long time without releasing any significant amount of CO_2, since the energy comes from nuclear fission rather than liberating carbon from deep within the earth.

At the same time, as disasters like Chernobyl and Fukushima made vividly clear, failures when it comes to nuclear power can have catastrophic consequences. And while we might hope that we've learned our lesson and that such accidents will not happen again, the costs of such a mistake are phenomenally high. The land around Chernobyl will be radioactive and unlivable for thousands of years. Even if another similar event seems unlikely, the potential outcome is so bad that those who are more precautionary in their attitude about risk may think it's not a smart gamble.

In addition, the war in Ukraine has demonstrated that accidents are not the only threat when it comes to nuclear power. In March of 2022, the invading Russian army took control of the Zaporizhzhia power plant, one of the largest nuclear facilities in the world and the single largest in Europe. Over the following months, the world watched with great concern as the risks from the reactor seemed to multiply. The most obvious threat came from the possibility of an errant missile or bomb hitting the facility itself, causing damage that would result in a failure. But that sort of direct assault is not the only worry. Russia's occupying army reduced the workforce of the plant, and concerns about human error caused by exhaustion or insufficient expertise took center stage. Finally, when fighting caused power disruptions to the facility, onlookers worried that external power failures could eventually lead to an inability to cool the reactor, resulting in a meltdown. In general, nuclear facilities in a war zone are a stark reminder of the risk that comes with harnessing nuclear power. Such considerations legitimize fear and anxiety around nuclear energy, and so even if a massive scale-up were the right thing to do from a climate perspective, it's unclear whether nations will be willing to take the risks of increased nuclear energy for the sake of the benefits of mitigating climate change.

Whereas the assumption of a nuclear power scale-up could be problematic if we don't have the political will to follow through, there is another assumption in the models used by the IPCC that raise a worry about realistic expectations. We have gotten to the point where the world simply cannot decarbonize quickly enough to avoid 1.5–2 degrees of warming only by reducing our emissions. What we need, in addition, is to take GHG out of the air. What we need, that is, is to capture carbon and trap it again, preferably in a long-term form it won't easily escape from. Of course, nature itself offers a form of carbon capture, as we can plant trees and rewild land in order to increase its capacity as a carbon sink; and this sort of intervention shouldn't be undervalued. But trees are both relatively short-term sinks (they'll die and rot and re-release their trapped carbon), and planting significant numbers takes a huge amount of land. Thus, scientists have turned to the hope of carbon capture that is decidedly more high-tech.

There are a couple of different methods being considered as negative-emissions technology. Although models care only that emissions are being absorbed somehow, and so there is no sense in which the climate models assume one version over another, most of the IPCC reports reference a particular technology—bioenergy with carbon capture and storage, or BECCS. This technology refers to the burning of biomass (think: fast-growing, highly photosynthetic plants) for energy while utilizing carbon-capture technology to capture CO_2 at the emissions point, like in a smokestack. Since the plants absorb CO_2 while growing, if we can capture it after burning, we can generate energy that is carbon-negative.

The major drawback to this plan is that it requires an enormous amount of land dedicated to growing the biomass, and so it is unclear whether it's really possible to do at the scale needed to help avoid dangerous warming. Scientists estimate that deployment of the kind assumed by IPCC models would require use of anywhere from 25 to 80 percent of the land currently used to grow crops. There are also technical issues regarding how efficient the entire lifecycle of a BECCS system can be made, including all the transportation of biomass, the capturing of the CO_2, and the sequestration (which may include yet more transportation). For all of these reasons, it's simply not clear that BECCS will save us.

The other category of negative-emissions technology avoids at least the land-use problem and is popular among the more techno-optimistic, as it does something that sounds borderline magical: it sucks CO_2 directly from the air and returns it to the ground. There's only one problem, which is that it doesn't yet exist in a scalable form.

What scientists have come up with so far are proof-of-concept models that show initial promise but that have very real drawbacks. Consider, for instance, Orca, which is the largest carbon-capture plant in the world, located in Iceland just outside Reykjavik. This facility is instructive for many reasons. First, despite being the largest plant in the world, it will only capture about 4,000 tons of CO_2 annually. And while that may sound like a large number, remember the scale that we're talking about: the all-time anthropogenic CO_2 budget is 3.67 trillion tons. On average, humanity emits more than 1,250 tons of CO_2 per second. What that means is that our largest plant to date, running around the clock for a year, will capture about 3 seconds' worth of humanity's emissions.

It's also very expensive. Experts predict that costs will eventually need to get to something like $100 to $150 per ton in order for this sort of direct air capture of carbon to be profitable without subsidies, but Orca's costs are between $600 and $800 per ton. The hope, of course, is that it becomes more affordable with scale, but there is no guarantee that the technology will ever be competitive, let alone in time to help us escape the immediate threat of warming.

Finally, the plant's location is important. Orca was built in

Iceland because it has two very important properties: plentiful, cheap geothermal energy to run the plant; and a particular geology that makes it possible to store captured carbon underground. While there will of course be experimentation with other locations and technologies, humanity's largest effort to date is only as efficient as it is due to the availability of geothermal energy, operating at a very small scale and at prohibitive cost.

Negative-emissions technology is nowhere close to being where it would need to be to save us. And it's possible it never will be. When we evaluate climate models that give us a chance of avoiding dangerous warming, it's important to note that they are basing that optimism on technology that faces many challenges, and which may never be able to play its hoped-for role.

None of this is news to the IPCC scientists, of course. While they want to provide a map of how humanity can yet avoid catastrophe, their warnings are getting more dire with every report. In the Sixth Assessment Report, the authors, for the first time, explore what it would take to limit warming to within the more ambitious limit of 1.5 degrees, rather than the original cap of 2 degrees. As one might reasonably guess from the foregoing summary, however, the models had a very difficult time finding a way to make that happen. The closest the IPCC gets to telling us that it's possible to do what we should and avoid 1.5 degrees of warming is in its evaluation of a lovely, optimistic future scenario in which the world's people and nations have a serious change of heart. Modelers characterize this scenario as "Taking the Green Road," and elaborate:

> The world shifts gradually, but pervasively, toward a
> more sustainable path, emphasizing more inclusive

development that respects perceived environmental boundaries. Management of the global commons slowly improves, educational and health investments accelerate the demographic transition, and the emphasis on economic growth shifts toward a broader emphasis on human well-being. Driven by an increasing commitment to achieving development goals, inequality is reduced both across and within countries. Consumption is oriented toward low material growth and lower resource and energy intensity.

Models that made all of the assumptions mentioned about the scale-up of nuclear power and the deployment of carbon capture, and which assumed that the world shifted in the direction of Taking the Green Road, weren't quite able to limit warming to 1.5 degrees, but they were able to overshoot it only slightly, and then bring the temperature back down by the end of the century.

So it's possible to limit warming to something in the neighborhood of what we've said is permissible. But it is simply not likely. When scientists recently attempted to quantify this judgment with computer simulations, they estimated that humanity has a 0.1 percent chance to avoid 1.5 degrees of warming in this century. We waited too long to take serious action, and we are clearly not prepared to do what is required.

I'm not saying this to be defeatist. I'm just being realistic. In all likelihood, the Earth will warm more than 1.5 degrees Celsius in the coming decades, which means we need to ask some important, difficult questions, such as: Where are we actually headed? And what will the world be like in that scenario?

———————

To know what the future climate will be like, we need to know about more than the physical sciences of the Earth; we need to know about people—how they'll react to an unfolding climate crisis, and what policies they'll adopt. Computer modelers can't predict "the" future. They can only tell us what we should expect given a particular human story, and it's up to us to write that story.

That's how we end up with something like the Green Road narrative. This is a "scenario," or a story about how the world looks, that gives modelers a reason for making various assumptions. For instance, in a world in which we take the Green Road, not only do humans change dramatically in their willingness to act aggressively to combat climate change but they also act to reduce inequality. Investment in education and health lowers fertility rates, resulting in a lower human population than we might otherwise have, which means there are fewer people using resources. Modelers start with scenarios like these and then use them to specify particular policies or interventions that modify certain variables. In the case of the fertility rate, the models running the Green Road scenario use the United Nations' "low" fertility projection, which predicts that the human population will peak midcentury at around 8.5 billion and then drop to below 7 billion by the year 2100. This is a far smaller human population than the UN's median estimate that we will reach nearly 11 billion by 2100, but it's a possible future. It's one way things could turn out if we focus on reducing inequality and providing all people with access to various basic goods while investing heavily in green technologies that benefit all. And the computer models tell us that if we did go this route, we could just barely overshoot 1.5 degrees of warming.

But of course it's likely that things won't turn out that way. So the IPCC considers four other scenarios, describing four very different worlds. The most realistic picture of the future is described by a scenario called "Middle of the Road":

> The world follows a path in which social, economic, and technological trends do not shift markedly from historical patterns. Development and income growth proceeds unevenly, with some countries making relatively good progress while others fall short of expectations. Global and national institutions work toward but make slow progress in achieving sustainable development goals. Environmental systems experience degradation, although there are some improvements and overall the intensity of resource and energy use declines. Global population growth is moderate and levels off in the second half of the century. Income inequality persists or improves only slowly and challenges to reducing vulnerability to societal and environmental changes remain.

In this scenario, we continue to plod along more or less as we have been doing, with no miraculous change of heart. To be clear: this is not a "business-as-usual" scenario. To follow this path, the global community would have to meaningfully act, which it has been slow to do. Nations would need to make progress against their climate pledges, with at least some meeting or exceeding their climate-mitigation goals. But even if we do that, the planet is likely to warm by 2.7 degrees Celsius over the rest of the century, with a possible range of 2.1 to 3.5 degrees.

In other words: slow, uneven progress is not enough. If we don't make a dramatic change in the very near future, we should expect warming of nearly double the lofty aspirations of the Paris Accords.

And of course, there are much more frightening scenarios. If we wholly fail to mitigate climate change and just continue ravaging the Earth, we could yet see well over 4 degrees of warming, with a range that extends up to 5.7 degrees. And while I don't know any climate scientist who thinks we will fail quite so dramatically as to bring about that world, there is another scenario that does have some elements that ring true. Consider the world described as "Regional Rivalry":

> A resurgent nationalism, concerns about competitiveness and security, and regional conflicts push countries to increasingly focus on domestic or, at most, regional issues. Policies shift over time to become increasingly oriented toward national and regional security issues. Countries focus on achieving energy and food security goals within their own regions at the expense of broader-based development. Investments in education and technological development decline. Economic development is slow, consumption is material-intensive, and inequalities persist or worsen over time. Population growth is low in industrialized and high in developing countries. A low international priority for addressing environmental concerns leads to strong environmental degradation in some regions.

I imagine many readers will recognize some of the current world reflected in this description, and while it seems unlikely that it's the best prediction of where we will go from here, the future may well be some combination of Middle of the Road and Regional Rivalry.

In the world of Regional Rivalry, we should expect 3.6 degrees of warming.

Based on the scenarios above, it seems overwhelmingly likely that the world will warm well over 1.5 degrees during the rest of this century. Indeed, the UN Environment Programme's 2022 Emissions Gap Report "finds that the international community is falling far short of the Paris goals, with no credible pathway to 1.5°C in place. Only an urgent system-wide transformation can avoid climate disaster." Worse, if we don't make truly radical changes in the very near future, we flirt with an amount of warming that we talk far too little about: 3 degrees Celsius, or potentially even more if the higher end of the modeling ranges turn out to be accurate.

My worries about comfortably living in Cyprus during the summer, about fire season out west, and about the increased risks of living in many coastal regions are informed by my fear that we are likely not on one of the most optimistic trajectories. But of course, those concerns were initially framed as pragmatic considerations. They are reasons for me to plan my life a certain way. The much more important considerations regarding the future of our climate are moral concerns about what will happen to folks who are far more vulnerable than I am, and who may not have the luxury of choosing to avoid coastal regions or who can't afford to move away from heat that is devastating to crops and sometimes even lethal to humans.

As the world passes 2 degrees of global average warming, among the most serious ethical features of our situation will be that these vulnerable populations will be hit (and already are being hit) first and worst by the ravages of climate change, and yet they have benefited the least from the development that caused it. In a world where we allow climate change to go unmitigated for far too long, people like me who have benefited extensively from the fossil-fuel development of the first world will plan our lives to avoid the harms of climate change while the world's poorest individuals will suffer, be displaced, and die from various climate disruptions.

Just how bad might it get? Here's a glimpse of only a few of the harms that may take place if we follow something like the Middle of the Road scenario—perhaps sprinkled with some elements of Regional Rivalry—and so warm by 3-plus degrees over the rest of the century.

Although we commonly talk about climate disruptions rather than simply about warming, it is important to understand how dangerous heat is. Humans have evolved to live in a very narrow ecological niche: too hot and we die; too cold and we die. Still, you might think that 3 degrees Celsius is too small a change to make a difference, but we have to remember that this is global average warming.

If the world warms by 3 degrees on average, it does not do so evenly. The air over land warms faster than the air over water, so the average increase that we experience will be significantly more. Further, certain geographic areas warm faster: the Arctic, in particular, warms much more quickly than the rest of the world, which is important because that's where many of the stable ice shelves are. So 3 degrees of warming is often 4 or 5 degrees where people live, and may be 7 or 8 degrees in the Arctic, resulting in ice-free Arctic summers by the middle of this century.

But even if the average temperature we experience is 4 degrees warmer, why is that such a big deal? Because moving the average doesn't just move the temperature on an average day; it moves the entire distribution of temperatures up. For any average temperature over time, extreme events, such as heat waves, occur. But when you change the average, events that used to be extreme become relatively common, and events that hardly ever happened become the new extreme. (If you're accustomed to thinking in terms of normal distributions, or "bell curves," imagine a bell curve of temperature, where the average temperature is in the middle. Now slide it 4 degrees to the right, and notice that the temperatures that were several standard deviations from the norm before are now in the much fatter part of the curve. That's the problem of changing average temperature.)

Practically, what this means is that normal heat waves of the past become part of everyday life, and the new extremes are characterized by heat that most people have never experienced. Scientists estimate that if we allow the Earth to warm as much as 3 degrees this century, many people living in tropical regions will be exposed to dangerously high heat indexes on most days of the typical year, and the sort of deadly heat wave that used to be rare will occur annually. If we were to blow further past our carbon budget and allow 4 degrees of warming, then deadly heat waves that should happen only once every 740 years would become the norm by the end of the century. And a new species of unprecedented heat waves, which should happen only every few million years, the kind that humans have probably never experienced, would become commonplace, especially in the tropics.

Yes, most days get hotter. But the especially hot days get extremely hot, and the previously extreme days become unprecedented.

Heat can make us sick, require that we hide indoors and use energy-intensive technology (if we have it), and even kill us (if we don't). In fact, according to a recent study, climate change is already responsible for an estimated 37 percent of the deaths from heat waves around the world, and that share will only go up as increased temperatures continue to turbocharge extreme heat.

Perhaps most terrifying of all is a study that tried to estimate just how many people climate change could kill with heat in the coming years if we don't significantly step up our response. The authors' finding? If we were to allow the world to warm by 4.1 degrees Celsius, the increased heat around the world would cause an estimated 83 million deaths by the year 2100.

Eighty-three million dead. Just from heat.

Of course, climate change will wreak far more havoc than the damage heat waves cause. Among the most oft-cited dangers of climate change is sea-level rise, which is a function of both glacial melt and thermal expansion as the water warms (that is, the water itself increases in volume as it heats up). Many small island nations like the Maldives, Tuvalu, Seychelles, the Dominican Republic, and the Bahamas are at risk from rising tides—including existential risk. The smallest and lowest-lying nations are unlikely to survive even 2 degrees of warming, let alone 3 or 4. These small nations are not the only ones threatened: larger island nations like Singapore face significant risk from the water, as do countries like Bangladesh, which is densely populated in its low-lying delta region.

While the threat from sea-level rise is most dramatic in the case of small islands that may become wholly uninhabitable, it will cause serious destruction in very many coastal areas. Recent data suggests that as many as 190 million people currently live on land projected to be under the high tide lines by 2100, with over 630 million occupying land that will be below flood levels by that year. While the globally privileged will be able to simply move in the coming decades, the most vulnerable will have nowhere to go as their homes and livelihoods slowly disappear into the water.

Higher water combines, then, with the increased frequency of intense storms to deal dangerous and often deadly blows to coastal regions. At my university, I'm often asked to give lectures on climate-change ethics, and from year to year, I simply update my slides on extreme weather events from one terrible storm in the news (Hurricane Katrina in New Orleans, Superstorm Sandy in the Northeast) to the next (Maria in Puerto Rico, Harvey in Houston). Down the street from where I live is the adorable historical district Old Ellicott City, which had the bad luck of getting battered by multiple "once-in-a-thousand-year" floods in the course of two years.

When you hear that the same small town got hit by yet another "once-in-a-thousand-year flood," you might start to think that the phrase has lost some of its meaning.

We could continue. The droughts. The wildfires. The potential for destabilizing the Atlantic Meridional Overturning Circulation (a weirdly technical aspect of the ocean's current, which, frankly, I wish I had never heard of). New infectious disease outbreaks as vectors like mosquitos and bats change their migratory patterns. The threat of increased armed conflict over declining food and fresh-water resources. Immigration crises caused by the loss and destabilization of millions of people's homes. The ongoing mass-extinction event that is ravaging biodiversity. The economic hardship coming as we realize that mitigating and adapting to climate change will cost us trillions of dollars. (Indeed, one projection estimates that, if we allow the Earth to warm by 3.7 degrees, it will cost a cumulative $551 trillion—an amount that, incidentally, is greater than the sum total of wealth in the world at the time of this writing.) Climate change is a global threat unlike anything experienced by anyone alive today.

That's the power of changing the global average temperature by a few degrees.

Although this book deals with climate change, this is not a book about climate change, so let the discussion to this point suffice in order to justify a simple point:

The stakes are very fucking high.

Our actions now will determine the amount of suffering for years to come. If you're anything like me, when you consider facts like those I just listed, you think: I should be doing much more than I am. Shit—I should be doing everything I can. Now. Yesterday.

But wait. If things are as bad as I say they are, why should we take the threat of climate change as calling us to action rather than as a reason to simply eat, drink, and be merry? After all, if the world is going to burn anyway, then we should enjoy it while it's more or less livable, for tomorrow we may die. Once we recognize how unrealistic the Green Road scenario is, we should accept that humans are simply not going to prevent dangerous climate change. We're much too late to do anything meaningful, so why spend our time fighting against inevitability?

The answer to this challenge is not to insist that we have a realistic chance to prevent more than 1.5 degrees Celsius of warming. Although it's technically true that we could still avoid overshooting it by much, I concede that it's strikingly unrealistic to believe we are going to do what's required. The much better response to the pessimist is this:

It can always get worse.

Climate change is not binary; it is not on or off. It is scalar, which is to say that 2 degrees would be much worse than 1.5; 3 would be much worse than 2; 4 would be utterly catastrophic; and every increment over that would usher in more of the stuff of nightmares. But the flip side of this increasing scale of terror is that every effort we make could prevent suffering. Delays in changing climate policy or any amount of watering down national pledges will consign real people to suffering and death.

Remember the study that projected 83 million people dying from heat by the year 2100 if we allow warming of 4.1 degrees? Well, there's an optimistic component to that research. According to the authors, if we manage to limit warming to 2.4 degrees—still far too much, but a fairly reasonable estimate of where we could land with serious, aggressive, global action—then we could reduce the number killed by heat from 83 million to just 9 million.

In other words: by limiting warming to a level that's merely "very bad" rather than "catastrophic," we could save 74 million people from being cooked to death.

This feature of climate change means we can't give up. It is the result of this combination of scalar badness, urgency, and the catastrophic nature of the threat that makes climate change feel morally overwhelming. How can we possibly live decent lives when we're contributing to such awfulness? If you have ever thought of or worried about climate change, you've probably had such a thought. And while perhaps not everyone has articulated an argument for why it feels so hard to be a good person at this moment, it's not hard to reconstruct the sort of reasoning that seems to put us all under serious moral pressure. Consider the following claims, and where they seem to lead.

1. Most of what we do causes emissions, and so contributes to climate change.
2. But climate change is already seriously harmful, and continuing to contribute to it will just make it worse.
3. So most of what we do implicates us in a system that will cause significant harm to other people.
4. And for those of us in wealthy nations, enjoying world-historic benefits of fossil-fuel-intensive development, our contributions to climate change are worsening a problem that will be disproportionately bad for the poorest and most vulnerable (whose own contributions to the problem are often minuscule fractions of our own).

This sort of reasoning can lead one to think that just existing in the modern world makes each of us morally responsible for the harm we are helping to bring about. And if that's right, then perhaps the only morally defensible life is one of immediately divesting as much as possible from environmentally expensive living and doing everything one can to bring about a better, safer world. Such an obligation—should it exist—would reach into every aspect of our lives, rendering moral judgment on nearly all our daily actions. After all, in modern society there is virtually no decision that is carbon-neutral. One's work, hobbies, relationships—all are likely to increase one's carbon footprint.

Not even the most personal, intimate decision escapes such a responsibility. As suggested by the modeling scenarios' use of population as one of the variables influencing climate change, the decision to have a child certainly has a carbon impact. Indeed, for many of us, having a child is the single most environmentally expensive thing we can do, as it creates a whole new person who may then go on to create new people, who may create new people, and so on. A totalizing focus on the moral urgency of climate change, then, might suggest that many of us should remain childless in pursuit of living a low-impact lifestyle.

This initial attempt to think through the individual ethics of responding to climate change seems straightforward but intimidating. If sound, such reasoning suggests that all of us are on the hook for minimizing our carbon footprint, and very many (most? virtually all?) of us are failing egregiously. If the argument is convincing, then we may be forced to simply accept that most of us are not very good people. On such a view, living a decent life in such catastrophic times is prohibitively difficult—not because we don't know what to do but because the demands are high, and so we do not manage to do what we should.

Some people do make a serious effort to live up to this sort of standard, though. Perhaps you know someone like this. I know a few. One way to read this first chapter is as providing a strong case that, despite the discomfort of such a conclusion, they're right. Understanding the climate science seems to imply that we are morally responsible for much more than we may want to accept.

Not everyone is so quick to accept a move from the badness and urgency of climate change to a conclusion about individual obligation, however. In recent years, it has become popular for moral philosophers and environmental activists alike to object to putting the onus on individual responsibility.

Why? That's the other half of the climate ethics puzzle.

CHAPTER 2

I Don't Make a Difference

The power of consumerism is that it renders us powerless. It traps us within a narrow circle of decision-making, in which we mistake insignificant choices between different varieties of destruction for effective change. It is, we must admit, a brilliant con.

—George Monbiot, "The Big Polluters' Masterstroke Was to Blame the Climate Crisis on You and Me"

I was an undergrad the first time I remember thinking about my own small contribution to environmental problems. At Hanover College, in rural Hanover, Indiana, I felt myself being molded into an aspiring philosopher and a wannabe intellectual. As I learned about moral philosophy and environmental ethics, I was easily convinced of the importance of environmental protection. And I remember wanting to be a bit more like my hippie friends, who found it obviously right to live low-impact, thoughtful lives. You know the type: they love nature, don't shower as much as everyone else (to conserve water?), often are vegetarian or vegan, both for

animal-welfare reasons and because animal protein is environmentally expensive, and they wear the same twelve-year-old
clothes for days on end.* These folks took arguments like the one
presented in the previous chapter as obvious and decisive, and they
seemed righteous to me. Which led me to be a bit sheepish about
the fact that I wasn't entirely sure whether I was required to live
such a life.

One afternoon, probably in 2002 or 2003, I was hanging out
with a few of these more environmentalist classmates, shooting
the shit after a philosophy lecture, and the case of recycling was
raised as an example of a small act that one is obviously morally
required to do. Although I shared the intuition that recycling is
good (and I in fact did—and do—conscientiously recycle), I was
bothered by the case. After a minute or so, as everyone else took
the example at face value, I finally asked, "But does it really matter
if any of us recycles?" I remember feeling like they were all staring
at me judgmentally, though that might have been in my head. I
continued, "I mean, I'm just one person on a planet of billions of
people. Surely I can't make a real difference, right?"

As sometimes happens in a teaching-focused liberal-arts college like Hanover, one of the professors had joined the group and
was listening intently. When no one immediately answered my
challenge, she just gave me a warm smile, sidled up beside me, and
said, "Well, now there are two of us."

It was a rhetorically elegant move by someone whom I liked and
respected, so I smiled and nodded as if I understood the depth of
what she had just conveyed. And the rest of the students continued

---

* Note from one such friend: "*Twelve*-year-old clothes? Fucking amateurs."

the conversation, thankful that the weird Earth-hater had been silenced.

But I was not satisfied.

I think the reason I have this memory—despite having virtually no context around it (including who was part of the group)—is because I'm still not satisfied. Two decades later, I'm still bothered by it. Despite shutting my mouth in the moment, I remember thinking to myself, Well that's not an answer at all. Because two people still isn't enough. And if we're joining a larger movement, then that movement would exist with or without me. So I'm still just bringing the recycling resources of one person.

Although I wouldn't discover it until much later, it was around this time that a philosophy professor in North Carolina was making rigorous my loose, inchoate worry. In 2005, Walter Sinnott-Armstrong, professor of philosophy at UNC–Chapel Hill, published a now-classic paper titled "It's Not My Fault," which investigated precisely the question of whether individuals have an obligation to take certain concrete environmental actions.

In particular, Sinnott-Armstrong investigated the case of needlessly emitting GHG by taking a Sunday joyride in an inefficient vehicle—or, as it would come to be known: "joyguzzling." Although he acknowledges that climate change is a terrible threat, and that GHG emitted by human activities is causing dangerous climate change, Sinnott-Armstrong argues that there is no obligation to refrain from joyguzzling. Joyguzzling is not wrong, despite many people's intuitions (including his, he records) to the contrary. The basic thought grounding this argument is the same one that started haunting me as a lowly undergrad in the context of recycling: a single action of a single individual simply does not make enough of a difference to justify treating it as an obligation.

Whether it's recycling or joyguzzling, my actions don't seem to matter.

Of course, some actions do make a causal difference; legislation by major national emitters, like the United States, can require drastic emissions reductions and so actually limit the amount of warming and climate disruptions. Even some private institutions, such as energy companies or multinational corporations in other sectors, can adopt policies that have significant reach and therefore materially reduce emissions. In these scenarios, it may well be the case that nations and corporations have strict moral obligations to reduce their emissions in order to prevent suffering. And some individuals, due to the political or social power they have, may look more like institutions in this respect than individuals: what they do can make a real difference. When Sen. Joe Manchin of West Virginia finally signed on to the Inflation Reduction Act in 2022 (which, despite the confusing name, is one of the most important pieces of climate legislation ever passed in the United States), that vote broke congressional gridlock and made possible an entire suite of environmental policies. Although it was the action of a single man, it was also a lever of policy. Such acts clearly have a real impact on the problem of climate change. But most of us, as private individuals, do not have such power. Our decisions to joyguzzle, recycle or compost, add solar panels to our home, take a transatlantic flight, or buy an electric car will not have a meaningful effect on the severity of the environmental challenges we face.

This sort of worry takes the argument of the previous chapter, accepts the science and moral urgency, but points to a gap in reasoning at the end. The environmentalist who feels morally compelled to divest from all GHG-intensive activities has drawn a conclusion about individual morality from a set of facts that impugn our collective morality. Climate change will be devastating if not addressed by the world's powers, and so they have an obligation to fix it.

What does that mean for each of us? Not much, so far. It might feel like each of us is on the hook for addressing climate change because we each emit GHG, but none of us can meaningfully address climate change by reducing our emissions. The problem is simply too big.

To put it more precisely, the structure that makes climate change so difficult to think about ethically is: it's a problem because of the harm it will cause; but none of our individual emissions cause any of the relevant harms. The first of these claims seems pretty straightforward, as the changing of the climate itself is not a problem. What makes climate change a catastrophe is that it will cause massive suffering. The second claim, however, is not obvious, so we should linger for a minute to understand why it's true.

Americans are some of the highest GHG emitters in the world, with the average of my compatriots spewing out somewhere around 15 metric tons of CO_2 annually. Although this is a lot compared to many in the world (for instance, the average Bangladeshi, facing huge risks from climate change, emits only 0.5 tons per year), it is an infinitesimal fraction of the total anthropogenic contribution. Recall the numbers from the previous chapter: to stay below 2 degrees of warming, humanity must emit less than 3.67 trillion tons of CO_2, which means that even the total lifetime contributions of a very wealthy, profligate person will not register in any noticeable way. To then focus on a single activity, like joyguzzling, further complicates the problem. Sinnott-Armstrong uses the estimate that a single joyguzzle might emit 14 kilograms of GHG; if that's right, then a claim that joyguzzling is wrong because it causes or worsens some harm requires that we believe that a few kilograms of CO_2 will meaningfully worsen a problem that arises only when trillions of tons of GHG collect in the atmosphere.

That just doesn't seem plausible. The sheer scale of climate change makes it insensitive to the sorts of normal, individual emissions that most of us create. But it's not only the scale that renders each of us causally impotent—it's also the complexity of the carbon cycle and the planet's climate systems.

Despite the initial gloss on the problem of scale, you might think that small contributions still matter. After all, when enough GHG is sitting in the atmosphere, every contribution counts, pushing us incrementally closer to disaster, right? Indeed, this seems to be an implication of something the IPCC says in one of its summaries for policymakers, with the dramatic subheading, "Every tonne of CO_2 emissions adds to global warming." While the scale of the problem makes it difficult to think about, we can compare it to a bathtub, full up to the rim, into which we are adding one droplet of water at a time. Eventually, the addition of a single drop will push some amount of water over the edge of the bathtub, despite the scale of the tub swamping the scale of individual droplets.

This sort of analogy has been very influential in thinking about the moral urgency of reducing GHG emissions, but it's also very misleading. The atmosphere is not a bathtub, and our individual emissions are not additive droplets.

The atmosphere is one layer in a massive and complex carbon system, through which carbon passes after it's liberated from solid forms and before it is absorbed into another layer. When we burn fossil fuels, we take carbon out of the ground and release it into the atmosphere in the form of carbon dioxide. But what happens next is radically uncertain. Sometimes that CO_2 is absorbed into the ocean, and sometimes it is absorbed by growing trees or other plants or absorbed into a swamp. All of these components make up the world's "carbon sinks" and are some of the ways carbon is removed from the atmosphere naturally. But some CO_2 stays in the atmosphere, and how long it stays is also wildly uncertain. Some molecules will stay for tens of years, and some for tens of thousands of years.

One layer of complexity, then, is that we don't know if the CO_2 we emit will contribute to climate change at all, in the literal sense of going into and staying in the atmosphere, where it infinitesimally increases the concentration of GHG. We can model the likelihood of our contributions playing a certain role, but the probabilistic nature of the carbon cycle seems to undermine a direct causal link between individual actions and any bad outcome. Though before we get too excited about the possibility that some of our emissions end up in natural carbon sinks, we must also wrestle with the fact that not all forms of carbon storage are equally good. When we release carbon from deep within the earth, we liberate it from a form that took millions of years to achieve and put it into a much more volatile part of the carbon cycle. Even if it is absorbed by trees, those trees will die and decompose, eventually releasing the gas back into the atmosphere. And if it's absorbed by the ocean, it contributes (again: infinitesimally) to the acidification of seawater, damaging fragile ecosystems. So some of what we don't know is, more precisely, where our individual contributions will go within the carbon cycle and thus how we ought to think about the likelihood of such small contributions.

It's even more complicated, though, because the particular climate harms don't happen in an obvious, predictable, linear way as a result of each individual contribution to GHG in the atmosphere. Rather, the climate harms are the result of global weather systems interacting in phenomenally complicated ways. Recall that the problem of heat is not, primarily, very small increases in average temperature; rather, the harms come from increases in extreme events, such as severe heat waves. The same is true of weather events like hurricanes, which increase in severity largely due to warmer water and drop more rain during individual events due to warmer air. Extreme events within a predicted average change, then, are probabilistic: every small increase in global average temperature brings increased risks of discrete events, and those events can cause harms. But we've now gone several steps from our already-very-tiny GHG contributions to any particular harms. Even if our molecules of CO_2 do stay in the atmosphere, and even if you can bring yourself to believe that such a small amount of GHG does meaningfully contribute to a minuscule fraction of warming, that warming isn't a harm. The harms are related probabilistically to the warming that you may not have meaningfully contributed to anyway.

Distressingly for those of us with environmentalist tendencies, it kind of looks like what you do just doesn't matter. The average, private citizen makes choices that result in emissions that are so infinitely small in comparison to the global carbon cycle, which is so infinitely complex, that there seems to be no straight line to be drawn between emitting action and climate outcome. And if what you do doesn't matter to the outcome, how could you be morally required to do any particular thing?

Joyguzzle away, apparently.

Climate change is a collective problem, and so it will be solved by collectives or not at all. Most individuals alone can't do anything meaningful to address the problem, so it's difficult to understand on what grounds one might be obligated to work to address it.

The case for an individual climate ethic might get even worse, though. Even if we can't make a significant difference to the harms of climate change through our actions, we might think that it's important to try to get lots of individuals to reduce their carbon footprints, since focusing on individuals might be a way of eventually getting large populations to reduce emissions. But this, too, has come under fire.

The excellent journalist and environmental writer Mary Annaïse Heglar provocatively titled one of her essays "I work in the Environmental Movement. I Don't Care If You Recycle." In it, she pointed out the meaninglessness of individual action in the face of climate change (akin to raking leaves on a windy day) but wasn't thereby saying that we shouldn't engage in individual climate action; rather, she was arguing that it is a distraction from what's important. Since solving climate change is a collective responsibility, Heglar warns us not to focus on individuals at the expense of institutions.

Many other contemporary thinkers agree, worrying about a sort of backfire possible if we focus on individuals rather than institutions. In a world where just 100 companies are responsible for

71 percent of total human emissions, it seems not only ineffective to focus on individuals but perverse. After all, if institutions are the problem, then we let them off the hook by focusing on individuals.

Perhaps most disturbing to some thinkers is that this sort of deflection of responsibility is likely what energy companies and other major corporations want: if we're pointing the finger at one another, we're not pointing it at them. Climate writer David Wallace-Wells calls the tactic of focusing on individual climate contributions a "neoliberal diversion" from the real culprits, and Heglar notes that the general model of making individuals feel responsible for their contributions in a system designed to force them to emit is akin to "victim-blaming."

Climate scientist Michael Mann brings together several of these points in a compelling picture of how the idea of individual responsibility for climate change can be (and has been) weaponized to prevent real climate progress. In an interview about his book *The New Climate War*, he says,

> Any time you are told a problem is your fault because you are not behaving responsibly, there is a good chance that you are being deflected from systemic solutions and policies. Blaming the individual is a tried and trusted playbook that we have seen in the past with other industries. In the 1970s, Coca Cola and the beverage industry did this very effectively to convince us we don't need regulations on waste disposal. Because of that we now have a global plastic crisis. The same tactics are evident in the gun lobby's motto, "guns don't kill people, people kill people," which is classic deflection. For a UK example look at BP, which gave us the world's

first individual carbon footprint calculator. Why did they do that? Because BP wanted us looking at our carbon footprint not theirs.

Energy companies and other bad or self-interested players love it when we, the populace, are asking one another if we have changed our lightbulbs, carry a reusable water bottle, or drive a fuel-efficient vehicle. Anything to keep us from talking about their role in torching the planet.

These arguments are all compelling, and they point out the risks of emphasizing individual climate ethics. In short, the risks are of two kinds: an empirical one and a moral one. The empirical risk predicts that focusing on individuals will lessen overall climate impact because it's a distraction. And the moral risk is that we hold the wrong parties accountable—although it's companies, politicians, and other structural players who are guilty in various ways and should be blamed, we both fail to hold them accountable and blame the victims.

If these arguments are sound, then not only might it be the case that individuals aren't obligated to make small changes to combat climate change but it's also wrong for us to focus on that topic in general. Indeed, perhaps it's wrong for me to write this book!

To say that we shouldn't focus on individuals in the context of climate, and that we should hold collectives accountable, sounds perfectly correct. But does that imply that there is *no* role for individual morality? Of course, I could be deluding myself, but I don't think so. And indeed, many of the folks concerned that we're pointing our fingers at individuals are concerned because they think it's wrong to do so. What we should be doing is focusing on the corporations and policymakers.

Heglar, for instance, says, "I don't care how green you are. I want you in the movement for climate justice . . . This is your planet, and no one can advocate for it like you can. No one can protect it like you can." Of course, you might not make much of a difference to the climate movement, just as you don't make much of a difference to emissions; but Heglar says that you can discharge your moral responsibility by advocating rather than by, say, composting.

That sounds like ethics to me. And it sounds like, despite the warning not to let the pressure off energy companies and politicians, there is good reason to wonder what each of us should do. I want to know how to live my life, and every day I have to make decisions and engage in actions that require environmental resources. But I also have to decide how to use my limited advocacy resources and where to spend my time and money. Is it better to march in the streets than to reduce one's consumption? Are we morally required to do both? Or neither?

Before we move on to building the tools to help us answer those questions, though, we have to make one more quick stop. So far, the arguments of these first two chapters have made it sound like climate change is unique—it's morally special, due to its massive size and complexity, which leads individuals to feel small and impotent. But is it really unique? Do no other problems have this structure?

CHAPTER 3

The Puzzle Is Everywhere

How can I begin to take action to discharge my responsibility in the face of such massive and diverse problems?

—Iris Marion Young, *Responsibility for Justice*

On March 7, 2020, I was in a Las Vegas ballroom full of spine surgeons, speaking to them about responsible opioid prescribing. It was a fantastic event, with a receptive audience and a gracious host, and none of us were thinking about the novel coronavirus that had already made its way into the United States. I had attended the conference chair's reception and dinner at the Cosmopolitan the evening before, and it had been packed. We stood around, shoulder to shoulder, leaning in and shouting at one another, the way you do—the way we used to do—in a loud, intimate setting.

Immediately after my talk, the conference host set me up at a table in the reception hall, where I signed books, shook hands, and listened to the stories of a long line of experienced physicians. I later half joked on Facebook that I immediately ran to the bathroom after it was over and nearly took a shower in the sink; but despite the urge to wash my hands, I hadn't yet really changed my behavior in response to the new infectious disease threat. No one at the conference had. We had no idea what was coming.

On March 10—two days after landing back home in Baltimore—I taught my last pre-pandemic class. That evening, the president of Johns Hopkins University sent a campus-wide email, informing us that in-person activities were canceled.

The era of the Covid-19 pandemic had begun.

Full of indecision, I waited another full day after Hopkins closed its doors before canceling a talk I was slated to give at the Hospital for Special Surgery in NYC that Friday. I couldn't help but worry that I was overreacting; and if I was, it was wrong of me to back out of a speaking obligation. But if this wasn't a drill—if Covid-19 was the real deal—I was absolutely obligated not to travel when I didn't have to. I should stay home, stay safe, keep my family safe, and help to "flatten the curve," as the rallying cry would soon be instructing us.

Of course, staying home that week was the right thing to do. NYC absolutely exploded with cases of Covid-19, becoming ground zero for the first wave of America's epidemic. And the entire country experienced something like the whiplash I was experiencing: going from business as usual to sheltering in place practically overnight.

The result of this whiplash was months of trying to figure out how to act in the face of a global pandemic, the likes of which no one had seen for a hundred years. And so there began a period of think piece after think piece in which epidemiologists, immunologists, journalists, science writers, and occasionally even philosophers debated what one ought to do at the various stages of the pandemic. Is it permissible to hire someone to grocery shop for you? Are you obligated to forgo luxury outings if going out would increase the chance of a local viral outbreak? How ought you to act in this strange new world?

These questions had many layers. In practical terms, many people wanted to know what they needed to do to keep themselves safe. There was so much we didn't know, so what sorts of risk was it rational for individuals to take? That question is largely not an ethical one. Although some moral philosophers believe we have duties to ourselves that might prevent a certain degree of risk-taking, taking a chance with one's own body, health, and even life for the sake of obtaining something they value is largely seen as permissible. That is, one is allowed to engage in risky activities, so long as they don't harm others.

That's where the ethics first comes in with infectious disease, though—because you might, in fact, harm others if you take risks with yourself. During that first pandemic spring break in 2020, college students were endlessly shamed in the media for flocking to the beaches of Florida, while others were held up as irresponsible for continuing to have parties or, when there were no lockdowns imposed, dining indoors or attending concerts. The first step of the moral argument against these activities was that people who took such risks were not only risking their own health but also the health of those with whom they came into contact. Early on, we learned that Covid-19 posed a risk of asymptomatic and pre-symptomatic spread, which meant that becoming infected could lead one to unknowingly share the virus with anyone else they visited. A common charge, then, was that young, healthy people who may not feel too at risk themselves were putting their parents or grandparents at risk by engaging in risky activities and then visiting those older, more vulnerable adults.

It's still unclear how risky certain activities were, and what level of risk it was permissible for younger, healthier people to impose on populations more vulnerable to the disease. Let's set that aside for the moment, however, because there was yet another moral claim being made, which continued throughout the pandemic: individuals had a duty to social distance, mask up, eventually get vaccinated, get boosted, and so on, in order to "slow the spread" of the virus or "flatten the curve" on the infection-rate graphics. And it was this feature that I saw tripping up people again and again. A requirement not to hurt someone else makes sense. We know how to think about that sort of rule, and most of us think it's justified. But if we have a further responsibility to contribute to a collective health goal, even when not at risk of directly harming another person, that's more complicated. What, precisely, does it mean that I have a duty to slow the spread of the disease?

One way to think about it is that the SARS-CoV-2 virus would stay around, hurting and killing more and more people, as long as it had a high infection rate. The R_0 (pronounced R-naught) of a virus is its infectiousness, and it is shorthand for how many people an infected person will then go on to infect. A virus with an R_0 higher than 1.0 will spread, while a virus with an R_0 less than 1.0 will start to die out; this is because as long as each infected person gives the virus to more than one other person, an outbreak grows—and the higher that R_0 value, the more quickly it grows.

The R_0 of a virus depends not only on its biology but also on social factors: social distancing and masks reduce the R_0, even for a highly infectious, airborne virus like SARS-CoV-2. So what does the imperative to help "slow the spread" mean? Well, it basically means to do one's part to lower the R_0 of the virus. Be part of a population that takes measures to reduce that reproductive number so that the virus spreads more slowly (and in the best case, with some diseases, eventually dies out, which is what happens when a population achieves herd immunity).

When the ethics of social distancing, or masking, or even vaccinating is put this way, a familiar structure starts to emerge: one is being asked to take an individual action in the name of a very large, structural good. I typically cannot, by myself, meaningfully lower the R_0 of a virus in my community; but each of my small, private actions can be part of what contributes to that effort.

In other words: Covid ethics starts to sound a bit like climate ethics.

If the structure of these two public health challenges is, in fact, similar, then we should expect people to be similarly confused in the two cases; and indeed, I think this is precisely what we find. Those think pieces written about the responsibility or duty of individuals, or about what we ought to do in particular, individual instances, often struggled to disambiguate the different sorts of reasons people have for not taking certain risks. Sometimes we want to chastise others for taking unnecessary risks for themselves or for risking harm to certain identifiable others. But sometimes our concern is more about whether they are holding up their end of society's burden. And it's not surprising at all that this last reason is more difficult to make salient to people. If their action doesn't make a meaningful difference to slowing the spread, are they really required to forgo the things they want?

We might just end up wondering whether any individual is ever obligated to social distance, mask, or even get vaccinated for the sake of some collective good like slowing the spread of a disease.

———

Climate change is so massive, and so complex, that most individual actions don't make a real difference to the problem, and yet it seems (to many of us) that individuals have some moral responsibility to do something about it. But as the Covid example makes clear, climate change is not the only challenge that has this structure. Indeed, as I've been working on what we can call "The Puzzle" of individual action in the face of collective threats, I've started to believe that it arises in many areas of our moral reality.

In a complex, globalized society of some 8 billion people acting in uncoordinated ways, the structure of the Puzzle is reproduced everywhere. That doesn't mean it will look precisely the same in every context, but I think we should expect to find many instances of a general problem that has the following properties: (1) a massive harm threatens, which makes the moral stakes feel high; but (2) individuals are largely powerless to affect meaningful change; and so (3) this leads to passionate disagreement about what individuals are morally required to do.

Although climate-change ethics has elevated the Puzzle to broader public consciousness, it was discussed fairly broadly in another ethics literature a bit earlier, in the moral case for vegetarianism. For many of us who have thought about it very hard at all, the case for not eating animals is about as strong as a moral argument can get. We raise and kill animals (often torturing them by raising them in nightmarish environments in the name of increased efficiency), causing massive suffering. Thus, eating meat is bad for the welfare of animals. In addition, animal agriculture is very bad for the planet. To take just the example of climate change: animals are a very inefficient source of calories, requiring a huge amount of energy to raise them and thus produce the food that arrives in supermarkets and on dinner tables. Recent estimates suggest that as much as 14.5 percent of total GHG emissions come from animal agriculture. That is not to mention, of course, that raising animals also requires significant land and water use that contribute to environmental problems. The practice of eating animals, then, causes massive suffering for animals that are quite capable of feeling pleasure and pain (they are what philosophers call "sentient" beings) and is a major driver of climate change, which will cause (and indeed, is causing) major suffering among humans as well.

What's the case for eating meat? If torment and environmental degradation are on one side of the scales, what are the reasons in favor of being a carnivore? Well, we like the taste of animal flesh. In the immortal words of snide meat-eaters everywhere, faced with the above argument by their vegan friends: "Sure, sure, sure. But bacon tastes good."

Ethically speaking, then, vegetarianism or veganism should be a no-brainer. Eating meat is bad for animals and the environment (and, truth be told, for many of us in the overdeveloped world with meat-rich diets, it's bad for us too), and the main reason we continue to do it is for gustatory pleasure. While our desires offer a good explanation of why we continue to eat meat, they are not a very good justification. Defending something that has numerous bad consequences by appeal to fairly modest desires makes us look selfish: we're simply not willing to work very hard at all to make the world a better place.

But would any individual becoming a vegetarian actually make the world a better place? Here, too, the Puzzle gets a foothold. For many of us living in societies with plastic-wrapped chicken breasts and ground beef in every supermarket, and meat dishes at every shared meal, catered event, or buffet, the idea that our individual choice matters to any bad outcome seems ludicrous. If I forgo buying the animal product at the store, will any producer notice and thereby torture or kill fewer animals in the future? Almost certainly not. And the case is even worse for the catered event: any meat dishes that aren't eaten will likely be thrown away. Is there any causal story possible to explain how eating meat in such a scenario contributes meaningfully to suffering?

Identifying the Puzzle in the context of meat-eating makes clear that our consumption choices in general can often face this problem. In buying something, we are rewarding the seller of that product with our business, which is why we are sometimes encouraged to boycott certain products or businesses if it is made public that they are unethical in some way. But if boycotting is a reasonable response to discovering bad behavior by a company, then in our globalized marketplace, we have a lot of work to do. There may be all sorts of horrors in the complex global supply chains that bring us our commodities, including the use of slave labor, child labor, or other exploitative labor practices (such as minuscule wages and/or inhumane hours), and those are only some of the most egregious examples. CEOs of large corporations may also be awful people—they may be racist, misogynistic, or hateful in other ways—and so supporting a company may seem objectionable because it supports a morally bankrupt CEO.

People are sometimes told that, in an open marketplace, they are supposed to "vote with their dollars," in which case purchasing is a powerful act of support. This makes good sense of the idea that we should boycott bad individuals or corporations, but it also quickly becomes incredibly demanding. Who among us knows all of what we are supporting on a daily, weekly, or monthly basis with our dollars? And how confident are you that, should you find out, there would be nothing objectionable uncovered? In the world of cheap food, experiences, and products, much of what we spend our money on is likely morally compromised in the sense that it supports some bad individual, corporation, or practice. Unless you spend an inordinate amount of time researching it, the chocolate you buy may well involve environmental destruction or exploitation, and the clothes and electronics you buy may well utilize child labor somewhere in the supply chain. Of course, all of it contributes to climate change just by being an instance of consumption.

Each of us, always, is contributing in multiple ways to many massive systems that cause harm—most of them without us even knowing it.

Climate change gives us the purest case of the Puzzle, but once the general structure is recognized, we can identify it elsewhere. When it comes to the climate, the scale is phenomenally large and complex, and at the level of the average private citizen, there are no individual actions that can meaningfully worsen the harms. In other contexts, the situation is less clear.

When it comes to animal agriculture, individual purchases of factory-farmed meat seem remarkably similar to individual emitting acts; any particular decision to forgo the purchase will not register causally, as the system is insensitive to individual choices. But of course, there are other ways to procure meat. Local farmers markets, small co-op farms, and even hunting are possibilities. And the calculus clearly changes in these contexts. If I buy meat from a local farmer, it might be more reasonable to assume that the business owners are sensitive to fairly small shifts in demand, of which I could be a part. In such a case, it might seem more likely that I'm making a difference. And if I am, then some classical moral reasoning might work for this case: I'm meaningfully contributing to harm for the mere sake of gustatory pleasure, so if that's not justifiable, the Puzzle might not arise in such a case. However, vegetarians and vegans are not typically asking only those who make a causal difference with their meat-eating to change their diet. They want animals to no longer be raised and killed for food, and that's not a state of affairs that any one of us can bring about.

The same is true of infectious disease. As noted earlier in this chapter, there are good reasons to protect oneself and particular others from diseases like Covid-19, which warrant actions like getting vaccinated and wearing a mask during different stages of the pandemic. But insofar as we call on people to act in the name of protecting large populations by "doing one's part to slow the spread of the virus," we impose the same sort of tension between the structural and the individual as arises in climate change: yes, we need to solve the problem, but I can't do it, so it's difficult to see why I would be morally required to do some tiny part that may not even register causally on the outcome.

In these cases, whether or not we have classical moral reasons to take individual actions, we also sometimes exhort one another to change our behavior for a collective reason. For many of us, it feels like we should act at least partially because we should be part of the solution.

Just as some moral philosophers and climate writers used the scale and complexity of climate change to argue that we should change our focus from individual moral responsibility for climate change to institutions and governments, similar themes have shown up in these other domains. Objections to vegetarian arguments (and responses to those objections) have often focused on whether individual meat-eaters causally contribute to harm. And in the wake of the Covid-19 pandemic, many of the same points were made about various interventions designed to slow the spread of the virus. When politicians or scientists began to shame individuals for contributing to the spread, others objected that the pandemic would not be solved by individualism. Collective problems will be solved collectively or not at all, and so in each case, there tends to be blowback from those who see a focus on the individual as a form of scapegoating or even victim-blaming.

Analogizing from Michael Mann's argument about climate, we might call this tendency to attack and blame individuals a distraction from discussing and demanding the policies that could actually work. Demanding that individuals go vegan lets the animal agricultural system off the hook. And asking individuals to mask and keep up on their vaccinations, but without policies that support those trying to stay healthy during an outbreak, is a form of individualizing public health that reliably fails to keep people safe while shifting responsibility for bad outcomes onto individuals.

If you are inclined to joyguzzle, I think you should probably stop. I also think that you should get vaccinated for infectious diseases like Covid-19 and the flu, not only to protect yourself and in particular others—but because you should do your part to protect your community. I think most people should eat few if any animals, and I think that if you find out that your favorite chocolate brand utilizes slave labor, you should boycott that brand by not buying it anymore.

I believe all of these things despite agreeing with much of the arguments laid out above—despite believing that in each case, you likely don't make a meaningful causal difference to the outcome, and for that reason they are problems appropriately solved by governments and other collective institutions. As a result of all of these beliefs, I agree that focusing on individuals can be a (successful) distraction, and so we must keep our eyes on the prize and hold our governments and institutions accountable.

That's why individual choice in modern times is a puzzle. It seems both to matter greatly and not to matter at all. The first step in solving the Puzzle is to recognize how it is often framed: Sinnott-Armstrong (of joyguzzling fame), for instance, makes clear that his argument is for a very precise claim: no one is obligated—or no one has a duty—to refrain from joyguzzling. This is because the traditional moral philosophies that are often deployed to explain the wrongness of an act don't work on the Puzzle.

Is he right about this? I think he probably is, but I also think—perhaps surprisingly—that it just doesn't matter very much whether or not we have an obligation not to joyguzzle. There are other moral tools that are often overlooked that help us to make sense of our intuitions when it comes to acting in the face of climate change, infectious disease outbreaks, and similar cases.

Laying out these moral tools—the ones that don't work in the case of climate change and some that just might—is the job of the next two parts of this book. It will involve a whirlwind tour of what I'll call "traditional moral philosophy," but in the service of demonstrating that there are other ways to think about ethics, which are needed to make sense of crucial parts of humanity's moral story.

Ultimately, we will require a new ethic, for a larger, more complicated world than there has ever been. We need Catastrophe Ethics.

PART II

ETHICS
STARTER KIT

CHAPTER 4

How We Try (and Often Fail) to Justify Our Actions

> It seems to me that the desire to be able to justify one's actions . . . is quite strong in most people. People are willing to go to considerable lengths, involving quite heavy sacrifices, in order to avoid admitting the unjustifiability of their actions.
>
> —T. M. Scanlon, *What We Owe to Each Other*

On the first day of the fall semester, I walked in late, having been unable to find my classroom (for some reason located in the basement of the engineering building). I apologized to the class, about half of whom were experiencing their first day of college, and then said that I would get to the roll call and syllabus shortly but that first I had a question for them.

"I want you to imagine a young pair of siblings—Mark and Julie—who vacation together in Europe one summer while on break from college." I paused to make sure they were paying attention. "During their trip, they stay one night at a particularly romantic beach house, and after some discussion, they decide to have sex. It was fully consensual, and just for fun. Even though Julie was on birth control, Mark also used a condom to make absolutely certain that she wouldn't get pregnant. The sex was good, they enjoyed themselves, they found it interesting, but they decided never to do it again and not to tell anyone. No one got hurt and, in fact, the experience brought them closer together. So, what do you think? Was it ethically permissible for them to have sex?"

I looked up from my notes to silence. And then a few giggles, while students looked around at one another, and then back at me, to see whether I was really expecting an answer. I leaned back in my chair, indicating that I was, indeed, waiting for an answer.

Ten seconds or so is a long time for there to be silence in a classroom, and so eventually, a student sheepishly raised his hand and, when I nodded to him, said—tentatively, as if walking into a trap—"Of course it's not OK. I mean, it's incest."

And that's the beginning of the demonstration. The story of Mark and Julie came from a study by social psychologist Jonathan Haidt and colleagues, who would give participants the prompt, asking if what Mark and Julie did was OK, then ask follow-up questions regarding their answer. I've informally run the experiment in many classes over the years, and especially with undergraduates, I am able to replicate the study results with pretty astonishing accuracy. It begins with the student who says that of course they were wrong, because they committed incest, after which I would say, "Well, yes, it's incest, but that's just redescribing the situation. What makes incestuous sex wrong in this case?"

In response, there's always someone in the class who is quick to offer that incestuous sex comes with a risk of causing genetic abnormalities in a child born as a result. But like the experimenter in

Haidt's study, I remind them that Mark and Julie use two forms of birth control. If the student persists in saying that the risk is still too high, I encourage them to do the math regarding the probability that a pregnancy with a genetic abnormality might result, and then consider whether every act with a very small risk of some bad outcome is morally prohibited. After all, we drive cars despite the risk of getting into a serious accident, because we judge the risk to be low enough to be justifiable.

Another student typically offers that the event would likely ruin Mark and Julie's relationship, and/or cause harm to the family if they should find out. So I remind them that Mark and Julie kept it a secret, and far from ruining their relationship, having sex brought them closer.

At this point, there is usually at least one person in the class who has become convinced by the details of the case, and they take over playing the advocate. They'll offer, "Well, if there were no bad outcomes, and they were consenting adults, then I don't think it was wrong. We might be uncomfortable with it, but that doesn't make it wrong for anyone else." And at that point, I can just call on people while my unwitting co-conspirator continues to push back against a classroom full of people who think Mark and Julie clearly messed up.

Finally, very often, someone in the class who is frustrated by the bizarro world they've apparently walked into will look around at the rest of us suspiciously and blurt out, "I don't know why, it's just wrong!" And that brings us to the conclusion of the narrative arc in the original study. Although the discussion can run shorter or longer, and with more or fewer detours into promising answers (more on those in a minute), at least some of the students tend to land in this place that psychologists call "moral dumbfounding"—holding on to an initial moral commitment despite the inability to provide reasons for it.

On Haidt's view, this study shows something important about people's ability to engage in moral reasoning: namely, that it's very limited. Not because incest is morally permissible but because subjects in the experiment so often remain committed to its wrongness even when they can't give any reasons for it. Similar to Joshua Greene's view of the brain as a camera with multiple settings, Haidt endorses a dual-process theory of cognition, in which there is a fast, intuitive, gut-response system, and a slow, deliberative, conscious system. The case of Mark and Julie demonstrates that our moral judgments are often the result of the fast, intuitive process, but we then do something strange—we backfill our reasons for that judgment with our slow, deliberative cognition after the fact. Haidt thus thinks that humans tend to post-hoc rationalize our moral judgments while believing that we're reasoning our way *to* those judgments. And perhaps surprisingly, we don't know this about ourselves. We may well sincerely believe that we're offering our reasons for coming to a certain view, even though we're coming up with them after having already made the initial judgment.

This summary of Haidt's experiment and the data it generated can sound quite pessimistic about the prospects for doing real moral reasoning. After all, if he's right, our brains just aren't built for finding our way carefully to a considered moral judgment, and that's a bit depressing. But there is an interesting note in his discussion of the incest study—almost an aside, appearing in just one short paragraph—which I think is very important. In this note, Haidt acknowledges that there is one group of subjects who tend to post-hoc rationalize less than others, and who will reason to a conclusion, endorsing whatever view is favored by a strong argument. That group? People trained in philosophy. These individuals, Haidt notes, are strange creatures, following reasons even to very bizarre conclusions (he calls out Socrates and Peter Singer as example weirdos, and so it's perhaps appropriate that we will meet both of them a bit later on).

Although this point about philosophy undermining our tendency to post-hoc rationalize goes by quickly in Haidt's reporting of the study results, I think it's crucial. And it's why I start many of my ethics classes with the case of Mark and Julie. The dual-process theory of cognition is, in a way, an argument for training in careful moral reasoning. Being morally dumbfounded means dogmatically sticking to a conclusion without reasons or evidence. Widespread, it's a powder keg for polarization and lack of genuine, sincere engagement with others. In short, a tendency to moral dumbfounding is a recipe for precisely the sort of world we see around us today.

If training in moral philosophy can help us to avoid moral dumbfounding and provide the sort of skill necessary for honest engagement on difficult issues, then it seems monumentally valuable. It's the prerequisite for debate and discussion that goes beyond table-thumping and tries to make real progress.

We should not, however, conclude from Haidt's study that you should endorse incestuous sex just because undergraduate students are often not able to fully articulate the ethical underpinnings of sexual ethics in a few minutes, without preparation. Commitment to careful reasoning and a healthy suspicion of our gut reactions does not mean that we must always find ourselves endorsing strange or even offensive conclusions. In fact, the sort of careful reasoning that might seem rare and a bit odd can not only help us to see when to abandon unjustified views, it can help us to understand the strong reasons for endorsing those views, even if we can't come up with them on the spot.

I've thought about the case of Mark and Julie for many years now, and I've decided that I think it's a bit of a trick. It asks us to judge whether they do something wrong in isolation, but that is not the only morally relevant question. Some of my students, if I let the discussion go on long enough, will identify the distinction between judging an act and judging a norm or policy. Perhaps what is most relevant is not whether these siblings acted wrongly in this one instance but whether we (most societies) should continue to endorse norms that rule out incestuous relationships.

On this higher-level question, the case is much easier to make. The power dynamics of a family make it very difficult for siblings to be on equal footing, and so we should be skeptical that the ideally consensual world of Mark and Julie would be commonplace. One sibling is likely older; in a patriarchal society, one being a man brings an unequal power dynamic; an entire history of growing together and knowing each other intimately allows for the possibility of psychological exploitation; and so on. A strong reason to condemn incestuous sex, then, is to protect the less powerful in a home from the more powerful, whether that's from an older sibling or a parent.

It's also relevant that sexual relationships are different from other forms of intimacy, and so having sex "on the table," so to speak, changes the dynamic of a relationship. Sibling relationships can be particularly valuable, and that value is born out of a unique form of intimacy. If incest weren't prohibited by societal norms or law, then we couldn't all rely on the knowledge that our family members are off-limits, sexually. This would dramatically change the kind of relationships that could develop within families, and the kind of safety and comfort one could count on.

There is very good reason to want children to grow up not seeing their siblings as potential sexual partners. We can thus endorse the norm that generally prohibits incestuous relationships, and still be less confident in what we should say about Mark and Julie. After all, for whatever reason, they did come to see each other as potential sex partners, and so now we have to decide what to think about their action given that fact. And it's not clear what to say. What I do feel confident saying, though, is that it would have been morally better if they had grown up without seeing a sexual relationship as a real possibility, and so there is strong justification for an anti-incest norm. And because the norm is justified, it may be correct to say that they were wrong, since their action violated a justified norm. But that judgment might not come with any need to blame or shame them, or to think that punishment is warranted. It's the norm that we are most concerned with, and not their (in this case, harmless) action.

The above argument is a case against incest that avoids moral dumbfounding. There are good reasons to prohibit Mark and Julie's action. So moral philosophy isn't always radical or corrupting. At its best, rather, it is clarifying. Exploring a difficult, personal, intimate question with transparency allows us to discuss it without resorting to table-thumping.

The case of Mark and Julie is my defense of some amount of philosophical skill. I think it's genuinely valuable—maybe truly necessary—to be able to reason well together in order to have a functioning society. But notice that my analysis of the case is not straightforward. I called the case "a trick," and I think many moral questions that we raise to one another are tricks in an important way—namely, that they lead us to think there should be a fairly simple, straightforward answer. My answer to Haidt's prompt is neither simple nor straightforward. I didn't answer whether it was morally permissible for them to have sex, but instead asked how we should think about the difference between justifying norms and individual actions. And at the end, the most important bit of most of our intuitions is salvaged by justifying a general prohibition of incestuous relationships, regardless of how we should evaluate the particular case of Mark and Julie.

This analysis is an example of one of my primary commitments in ethics, and this gets a little technical and controversial, so stay with me. But here it is.

Ethics is hard.

It's often very difficult to determine what we should do and how we should live, and yet we do feel the pressure to have an answer because, well, we have to live our lives. We make ethical judgments every day, and it would be incredibly uncomfortable to admit that we don't know whether they're the right ones, so we develop commitments, endorse them strongly, and treat them as if they were obvious. And then often, if Haidt is right, we likely rationalize them after the fact. So ethics is hard, but psychological pressures push us to treat ethical reasoning as if it were easy.

When I raised the Puzzle at the beginning of this book, you probably had an intuition about how to solve it. In fact, Haidt's experiment predicts that you likely jumped immediately to a judgment, and even if you had never considered it before, would defend that judgment if pushed. It might even feel like the answer is obvious.

In the rest of this book, I'm going to try to convince you that ethics is complex and nuanced—even if it feels obvious—because recognizing the deep complexity of various ethical issues can better prepare us to address the Puzzle. Facing ethical complexity is uncomfortable, and it feels good to have clear answers to ethical problems, so the most important part of developing moral reasoning skills is to reject seductive, easy, but ultimately wrong methods for thinking about ethics. Two of the most common seductive methods involve opposite inclinations—one tells everyone what to do while the other refuses to tell anyone what to do. Both methods are traps because they appear to solve our moral problems without actually doing so. A third method is popular with professional philosophers, and so is certainly not born out of a lack of training or practice, but it, too, is a trap. Once we learn how to avoid all of these traps we can go about the business of trying to build an ethics toolbox—one that is complex enough to handle tough cases and that therefore will give us a leg up in trying to figure out how to live decent lives amidst collective crisis.

CHAPTER 5

Trap #1: The Right Action Is the One Commanded by God

"But what will become of men then," I asked him, "without God and immortal life? All things are permitted then, they can do what they like?"
—Fyodor Dostoevsky, *The Brothers Karamazov*

I left the church when I was around fifteen years old, and I think it deeply shook my parents. Mom and Dad are both religious, in their own ways, so I'm sure my departure from organized religion felt a bit like a rebuke. And I'm sure they were worried about me—about my spiritual journey, about my soul, maybe even my virtue. To this day, they'll occasionally make comments to me about my atheism, showing how deeply it has colored their perception of me. But for my part, I think about it virtually not at all. I don't even really think of myself as an atheist, as that implies far more concern for a particular position than I have.

I just don't care about the God question very much.

I grew up Methodist in Greenfield, Indiana (no, no one else outside of central Indiana has heard of it either), and I can now identify some of my experiences with the church as deeply problematic. I absorbed patriarchal messages on the proper place of men and women—this from friends and acquaintances who presented themselves as deeply religious—and I even remember hearing a homophobic joke spoken in church by a church official. After having felt somewhat alienated from organized religion for more than a year already, the joke was the last straw. I went home that Sunday and told my parents that I wouldn't be going to church anymore.

Part of the reason I left the church, then, wasn't just because I stopped believing in God, though I was getting there; I was starting to think that church could be bad, spreading messages that were wrong but unquestioned.

I spent a lot of time in church, volunteering to work the sound system during multiple services a week. I alleviated the boredom of hearing repeat sermons by reading the Bible, sometimes fact-checking the pastor in real time. I became a big fan of Jesus Christ, because his messages so often struck me as powerfully correct. And I often took Jesus's central messages to be at odds with what I was hearing in sermons. Love thy neighbor, show warmth and compassion to the most marginalized in society, don't be over-reliant on material possessions (after all, it is easier for a camel to fit through the eye of a needle than for a rich person to enter the kingdom of heaven)—I could get behind all of that. But it never seemed to be the cornerstone of what the church taught my family and friends, gathered on Sunday morning.

I wasn't sure there was anything like a personal God to be concerned about, and I was becoming more confident in my own judgments about the world than I was in those being taught by religious authority—especially when it came to ethics. If

Christianity really did condemn homosexuality, then so much the worse for Christianity. I'm far more confident in my own judgment that there's nothing morally wrong with homosexual sex than I am that there exists a personal God who has judged it immoral. And on the flip side: I can admire the life of Jesus without believing that his example is important because he was the son of God.

However much this reasoning may have felt clear and comforting to me, it did not assuage my parents. My mom, in particular, worried about my departure from the church, and we talked about it regularly for years. In the style of good moms everywhere, she supported me regardless of my decision, but I could see that it ate at her. On returning home from college, I'd talk to her excitedly about my philosophy and religion courses and how they were helping me to understand my early, immature attempts to think carefully about God. And she would mostly listen but occasionally ask (sometimes distressed) questions.

She had at least two major worries. The first one led to a long, impassioned discussion one Christmas Eve. She and my big sister had been teasing me about whether I would burst into flames if I went with them to the midnight Christmas service (which I always attended, because it made Mom happy), and this had somehow led to a serious debate about the importance of God for morality. Mom and Sis were trying to convince me that God was in some sense morally necessary, because without Him, there was no divine reward for a good life or punishment for a bad life. At one point, my sister said something to the effect of, "If there is no God, no heaven or hell, then when Hitler died, the same thing happened to him as happens to anyone else!" And that seems, well, deeply unfair. Without cosmic reward and punishment, everyone gets away with everything, so to speak, when they die.

Further, my mom argued, that would mean that no one has any particularly powerful reason to be good. If rapists and murderers knew there was no afterlife in which they'd be punished, then they would be free to pursue their evil plans as long as they thought the benefits to them were worth the risk of the merely earthly punishments society might dole out.

At the time they raised the objections, I didn't have a lot to say to them, except that wishful thinking didn't constitute an argument. It sounded like they weren't giving any evidence that God, heaven, and hell existed but rather that it would be very bad if they didn't. And that's not an argument. As one of my philosophy professors liked to say: the fact that something would suck if it were true doesn't make it false (he called this the "wouldn't it suck fallacy").

There is, however, much more to say here. The challenge my mom and sister raised to me is called the problem of moral motivation; or, more simply, it is the question: Why be good? And despite my flippant reliance on the "wouldn't it suck fallacy," it stuck with me for years. The problem of moral motivation is hard—for everyone, not just for atheists.

My mom's second challenge took longer for me to understand, but it cut perhaps even deeper. She wanted to know: Who makes the moral rules if there is no divine rule-maker? Where does ethics come from, if not from the command of a being with divine authority and the ability to reward and punish?

A particularly famous version of this challenge is the one raised by Dostoevsky in *The Brothers Karamazov*, when Dmitri suggests that the nonexistence of God and an afterlife means that there are no moral prohibitions—or in Jean-Paul Sartre's punchy summary: "If God did not exist, everything would be permitted." For what force other than the command of an all-powerful God could prohibit an action? The truth of atheism, then, would imply moral nihilism, or the view that there are no moral facts. Although we might believe that there are, we're just wrong about that. We can call this second question from my mom the problem of moral truth: while the problem of moral motivation suggests that we need God to explain why we should be good, the problem of moral truth holds that we need God to explain how there even is good and bad, right and wrong.

After years of wrestling with these questions (thanks, Mom!), I eventually came to the view that they don't actually tell us much about the relationship between God and ethics. This is because, despite initial appearances, the existence of God wouldn't solve either challenge. I still think it's hard to explain why you should be good or how it is that morality exists; but it wouldn't be any easier if we could confidently appeal to God.

The problem of moral motivation is very old. It was one of the questions that Plato wrestled with more than two thousand years ago in his magnum opus, *The Republic*. Plato wrote in dialogue form, telling stories that typically involved his teacher, Socrates, addressing various interlocutors concerning important philosophical topics. Socrates, then, would provide the philosophical insight of the text (or so Plato seemed to think), while the conversational partners would prod, cajole, question, and often simply praise Socrates for his wisdom. (It's a running joke among philosophers that if you take Socrates out of a Platonic dialogue, all you're left with are his friends saying, "But of course! . . . How could it be otherwise? . . . By Zeus, it is just as you say!")

In *The Republic*, a group of Socrates's friends abduct him for a long evening of discussion, demanding that he teach them what justice is. It's a long treatise, consisting of ten "books," each of them providing plentiful philosophical insight on its own, but together constituting the most sustained argument we have from Plato. And among the topics covered over the course of the investigation into justice is the question of why one should be just. Socrates's friends Glaucon and Adeimantus note that pursuing justice can be personally costly and can even lead to horrible suffering. They thus demand to know if the life of a just person is worth pursuing, even in such a case. Why be good if it results in misery or torture, when evil people get away with a life of ease and pleasure? Even Socrates seems reeled in by the premise of their question, eventually putting forward a view on which the perfectly just person is the happiest and so the most well off, even if it doesn't look like it from the outside. (If that seems implausible to you, you're not alone; Glaucon started off quite suspicious too.)

I now recognize my mom's question as having the same sort of urgency as that of Socrates's friends. With religion, we can explain why you are better off being just. God can provide an answer, if He offers eternal reward and punishment for one's earthly deeds.

The problem is that it only looks like God provides an answer to moral motivation, because the assumption of divine authority actually raises another challenge, which is equally difficult. If you are trying to be a good person only in pursuit of eternal reward or fear of eternal damnation, you are failing at a core task of morality. This is because actions that are done for such a reason are deeply selfish, in that they aim at the long-term good of oneself. And being selfish seems at odds with morality.

The broader problem here is brilliantly laid out by Harvard philosopher T. M. Scanlon, who explores the problem of moral motivation in his 1998 book, *What We Owe to Each Other*. Scanlon tells us that answers to the question "Why be good?" face a difficult challenge, which he names Prichard's Dilemma, after the philosopher H. A. Prichard. According to Prichard's Dilemma, reasons to be moral cannot be "implausibly external incentives." That is, part of the value of moral action is that we do things for the right reason. Acting rightly because we fear God's punishment is the wrong kind of reason to be good. If I want to live a genuinely virtuous life, I must do the right thing for the right reason.

This first half of the argument strikes me as decisive. Although God seems like a good answer to moral motivation—explaining why Hitler should have behaved better and why I must do the right thing—it isn't. Because Hitler would still have been a monster if he had refrained from exterminating Jews and other minorities only because he was afraid of divine punishment. Sure, it would have been better if he didn't do evil; but what he should have done is refrain from evil for the right reasons—because other people have dignity, are worthy of consideration, their happiness matters, or any number of other good reasons.

Scanlon wasn't done, however. A dilemma is a challenge with two parts—"di-lemma," or double proposition—and so philosophers often present dilemmas to show how two different proposals are both problematic. This is how Scanlon employs Prichard's Dilemma.

On one "horn" of the dilemma (as philosophers like to say) is the requirement that answers to the challenge of moral motivation must not offer implausibly external incentives. So it would not seem right that you should be good in order to save your soul, or to become rich and powerful, and so on. But what's on the other horn? Well, philosopher Immanuel Kant famously argued that you should do the right thing simply because it is your duty. And while that has some intuitive pull, since it's actually connected to morality— you should act for the right reason—it turns out not to offer any real motivation at all. If my mom asks, as Glaucon did, "Why should I be good?" it seems completely unhelpful to answer, "Because morality demands it," or "Because it is your duty." After all, she wanted to know why she should do what morality demands.

This is the force of Prichard's Dilemma, and it reveals that answering the challenge of moral motivation is phenomenally difficult. Although there have been numerous solutions put forward, none of them enjoy anything like widespread acceptance. So I don't have a lot of confidence in any answer to the question "Why be good?" But I am confident that it undermines the religious answer: God's threats or promises are not the right sort of reason to explain why any of us should be moral.

In 399 BCE, Socrates had business with the courts of Athens. This was unusual for the philosopher, who preferred to spend his time discussing important philosophical matters, especially with the young men of the city, which is why his acquaintance Euthyphro was so surprised to see him just outside King Archon's court. King Archon was one of the Athenian magistrates, charged with overseeing religious worship and rituals, and so had legal oversight of charges involving offenses against the gods. Since Socrates was not one to get involved in such matters, Euthyphro asked him what he was doing there.

Socrates's answer is the beginning of a story that will end very badly for him. He has been accused by a young man named Meletus of worshipping false gods and corrupting the youth of Athens, so he is at the court on official business. These are the charges that will, in the coming weeks, result in the trial and conviction of Socrates—as punishment for which he will eventually drink hemlock and die.

But on the day he meets Euthyphro, he's still talking about how he should best respond to the charges. Their chance encounter provides him with an opportunity, as it turns out that Euthyphro is there to prosecute someone for wrongdoing—for being "impious," or "ungodly." Upon hearing this, Socrates notes that Euthyphro must therefore know what it means to be pious, and so implores him to share his knowledge, that it might aid in his own defense against the coming charges. Socrates becomes yet more convinced that he can learn from Euthyphro when he finds out the details of his case. Euthyphro is prosecuting his own father for killing a slave—a slave who himself was a murderer. Socrates is floored and begs Euthyphro to teach him.

This is the setup for another of Plato's Socratic dialogues, cleverly titled "Euthyphro." Whereas *The Republic* was primarily an investigation into the nature of justice, "Euthyphro" is an investigation into the nature of piousness, or godliness. Socrates claims to want to learn from Euthyphro, but—as is often the case in these dialogues—he asks questions that seem to show that Euthyphro doesn't, in fact, know what he's talking about. (This is basically what Socrates did with his life: going around, asking powerful people their opinions, and then making them look stupid, often in front of a crowd. It's not all that mysterious why the Athenians wanted to kill him; more mysterious, perhaps, is why it took them so long.)

So Socrates asks Euthyphro: You must be so confident that you know what is pious if you are to prosecute your father for murder, so will you teach me? And Euthyphro, clearly a man who prides himself on having such knowledge, happily obliges, trying out several possible accounts of piety over the course of the dialogue. Although none of his answers are ultimately satisfactory, one of them is the most famous, which is that the pious is what is loved by the gods.

In response, Socrates innocently asks, "Is the pious being loved by the gods because it is pious, or is it pious because it is being loved by the gods?"

With this simple-sounding question, Socrates launched more than two millennia of discussion and debate about the relationship between divine authority and morality, and it is the question that ultimately led me to believe that investigating God isn't necessary for investigating ethics.

Let's go back to my mom's second question about the role of God in ethics. It was, in essence, the worry that "if God doesn't exist, everything is permitted." According to her argument, God is necessary not only for having a reason to be good (that was the challenge of moral motivation) but for there even being a good. If there is morality, God must have made it. And if this is true, then studying religion is important, as that's the way to moral truth.

Socrates's question to Euthyphro challenges this view by offering another dilemma, now referred to as—you guessed it— Euthyphro's Dilemma. To make things easier, let's update Socrates's language a bit, and restate the challenge. For those who hold that religion is important because God tells us what's right or wrong, we can ask: If God commands an act (like love thy neighbor), is He commanding it because it is the right thing to do? Or is it the right thing to do because He commands it?

Socrates is being clever with the mirrored language, but his question is a simple one. If God commands something, that could be because God is responding to a moral reality that already exists (He commands it because it is right); or He could be bringing the moral reality into existence by the command (it is right because He commands it).

It might be tempting to say that God brings morality into existence by commanding it, since it is commonly thought that God is all-powerful and the creator of everything. But the problem here is that such a view makes ethics totally arbitrary. If God had commanded rape and murder, those would be right or obligatory, and refraining from them would be wrong. But this seems problematic on two counts.

First, if God makes something right by commanding it, then it means God isn't actually good in an important sense. God would be equally good if He commanded that people love their neighbors as He would be if He commanded that people torture their neighbors. It's part of our sense of God's goodness that we believe that God must be getting something right in identifying love and compassion as good. If God is good, His commands can't be arbitrary.

Second, there's a practical problem with the view that God creates morality by command, which is that God's commands are—at least in some cases—tested by our independent moral convictions. If someone tells me that God commands murder, I won't believe them. I'll distance myself from them. Even if God Himself came to me and commanded me to kill my son, I would not assume, as Abraham did, that I should therefore line up Isaac for sacrifice; my assumption would be that I had begun hallucinating and should take it very seriously as a symptom of some undiagnosed health condition. In the face of any claim that goes against my clearest, most deeply held moral intuitions, I will trust in my own moral reasoning. Indeed, this is part of what drove me away from church; when people in my life tried to convince me that homosexuality went against God's commands, I became immediately suspicious of their version of God. There must be a reason for caring morally about someone's sexual preferences, or else the command doesn't make sense.

So we arrive at the other horn of the dilemma: God must command things because they are right. This seems correct, and so we can answer Socrates's question to Euthyphro: the gods love the pious because it is pious.* The problem is that we now realize that Euthyphro hasn't actually told us what it is to be pious, or good. God is responding to morality, which is what we should do too. But that means we don't need God in order to explain morality, and so there is no reason to suppose that religion is the right path to moral knowledge.

---

*Logic-nerd alert: So we've seen two ways in which a dilemma can be used. The first way, demonstrated by Prichard's Dilemma, is to show that there are two ways to answer some question and they're both problematic. In such a case, we might have to go back to the problem or claim in question and revisit it. This is the reason I'm not really sure why we should be good—both solutions are trouble. Another way to use a dilemma is how I just employed Euthyphro's Dilemma: show that there are two ways to understand a view and show that one of them is problematic. This is evidence in favor of the other one. Dilemmas are powerful logical constructions, but it is common to respond to them by saying, "Hey, you claimed there were two possibilities, and tried to argue from there; but there are actually more than two possibilities. So, you know, not so fast!" If the objector is correct, then the original argument was a *false dilemma*, which is a common logical fallacy.

Of course, one might argue that God's commands are a short-cut to the truth, or the most reliable path to the truth, or something of that sort. But that doesn't seem true. Religious messages have often seemed wrong and, further, at odds with one another, with some religions or sects directly contradicting moral claims of others. Why should we think any one religion is most reliable? Determining reliability would require calibrating any religion's commands against our own views of what is right and wrong, which means we still have to do the work. There is no shortcut.

Euthyphro teaches us that we need to discover what is right and wrong, good and bad, for ourselves. If there is an all-good God, then doing the project well will mean we end up following God's commands. But that's because we're both—God and us mere mortals—responding appropriately to ethical requirements and norms that exist independently. Unsurprisingly, given Socrates's mission to goad people into doing philosophy, the lesson of Euthyphro's Dilemma is that we have to reason about ethics, whether God commands us to do anything or not. We can't escape our responsibility to determine what is good.

Appealing to God in the face of a complex ethical challenge is a trap because it would be nice and comforting if it worked, but it doesn't. I'm not arguing that God doesn't exist, or even that any particular religious teachings are wrong; instead, I'm arguing that even if God exists and if some religion or other is right, we still must determine why any given religiously justified claims are right for ourselves. Moral commands that don't make any sense outside of divine command shouldn't be trusted, because we have direct access to ethically relevant information. I know when an action hurts someone or violates their rights. I know an asshole when I see one. And we can talk to one another and reason with one another about these features of the world without bringing God into it. None of that tells us much about the nature of God. It just means that, whatever is true about religion, we can (and must) engage in ethics without referring back to it. And this is a good thing, given the deep pluralism of the modern world. If we had to determine the ultimate truth about religion before making any progress in ethics, that would be an insurmountable obstacle. Clear moral reasoning is our shared set of tools that allows us to sidestep that roadblock.

Returning to the Puzzle of individual morality in the face of collective problems, God won't solve that problem any more than He would solve any others. To be sure, some might claim that religious beliefs solve the Puzzle for us. In the case of climate change, one might say that we should limit our individual emitting behaviors because God commands us to be good stewards of nature. But someone else might claim that God gave people dominion over the Earth, and commands us to be fruitful and multiply, which they interpret to mean that we should use any available resources to be productive. Although I don't know whether God exists, and if He does, which is the more plausible command, I do believe that the right response is to try to reason about why God would command one thing rather than the other. What is the value of being an individual steward of nature, or of being productive? If religious folks are right, then the goal of moral reasoning is to be more God-like—not because we're blindly following commands but because we are responding to the same moral reality that He is.

Like in the case of moral motivation, I haven't solved what I called the problem of moral truth. Socrates didn't tell us (at least, not in "Euthyphro") what he thinks makes something the right thing to do; he just showed us that it's not divine command. The gods don't create morality; they recognize it, which is our job too.

What I think this means is that Dmitri's claim in *The Brothers Karamazov* is wrong: whether or not God exists doesn't matter much to the questions of ethics. There is certainly still a question of what is the right thing to do; it just turns out that God doesn't answer that question for us. That's why I said earlier that I don't really think of myself as an atheist, as that implies having a real stake in my view. In truth, I just don't think about God or religion very much. What I care about is living a good life, so that's what I spend my time and energy thinking about.

We must reason about ethics together, appealing to features of the world available to everyone. If God exists and commands certain actions because they are the right thing to do, then we will end in agreement, as we're identifying the same features of the world. Once we recognize that we must discover ethics for ourselves, however, a new challenge arises. God promised to answer our ethical questions with absolutism and universality: if God dictates right and wrong, then we should expect it to be clear and apply to everyone. But if God doesn't dictate morality, and so we have to discover right and wrong for ourselves, then perhaps something very close to the opposite is true. Perhaps, that is, morality is just whatever some individual or group says it is.

CHAPTER 6

Trap #2: The Right Action Is Relative

That's just, like, your opinion, man.

—Ethan and Joel Coen, *The Big Lebowski*

The idea that ethics comes from divine command is comforting because it claims to tell us straightforwardly what we are obligated to do. I think this comfort is behind a lot of people's insistence that they don't need to further investigate the ethical commitments they've been handed by religion. But there is another—very different—source of comfort that I've noticed among people when discussing ethics, which is the belief that there is no universally right or wrong answer to moral questions. There's just what's right for me.

I've noticed more of this latter view in recent years, especially among young folks who don't want to pass judgment on other people. Many of them have been raised in a cosmopolitan way, recognizing that people and cultures are different, and the value of tolerance has been impressed upon them. *You must not judge others, because you don't understand their culture or way of life.* Applying one's own belief to another's situation is seen as ethnocentric and closed-minded. But as a result, many of these folks overshoot tolerance and land on full-blown relativism. Whereas tolerance requires that we treat others with respect, even if we disagree with them, relativism holds that there is no real disagreement between those who hold different moral beliefs. Morality is relative to individuals or cultures, and so there is what's "true for me" and what's "true for you." Claiming that it's wrong, then—simply wrong, objectively wrong, universally wrong—is inappropriate, since there is no objective morality to help us decide between differing views.

To paraphrase the Dude, your moral beliefs are just, like, your opinion, man. (If that reference is lost on you, I highly recommend taking a break from this book and watching *The Big Lebowski*.)

The first thing to note is that there is no obvious connection between relativism and the toleration of others' individual and cultural beliefs. You can believe that others are wrong to practice according to certain moral norms and yet treat them and their views with respect (if their views are worthy of respect). And you can believe that there is no objective moral truth and yet treat other people like garbage for disagreeing with you.

In fact, there's a sort of paradox of tolerance-inspired relativism, which is that, if relativism is true, then tolerance is not the objectively correct stance to adopt. Those who feel pushed to relativism because it feels wrong to criticize other cultural beliefs often find it uncomfortable when I ask if another culture could be right in rejecting tolerance. If they are consistent, they must admit that it's possible. Relativism doesn't imply universal tolerance as a value; rather, it undermines it, by making the value of tolerance—like any other value—dependent on individual or cultural beliefs and norms.

I don't think many people are actually relativists. The view is attractive if you don't think very hard about it, because it sounds cosmopolitan and makes sense of the widespread disagreement about ethics. People disagree because they have different truths. Problem solved. But if we think a bit more about it, we have to ask some pretty uncomfortable questions.

Consider one such question: Is it possible that it's morally permissible to light a baby on fire for fun? If a society decides this makes great entertainment, could we accept that this is morally OK (for them)?

I take it that most people, however relativist their tendencies, have a very hard time saying yes. In my informal methodology of studying anyone who will have strange conversations with me, very few people are willing to out themselves as psychopaths in response to this question. And the reason is simple: while it's true that there is wide-ranging disagreement over moral issues, not all of that disagreement seems reasonable. At least sometimes, people and cultures seem to get the answer wrong. All of us who are opposed to lighting babies on fire for fun got this one right.

Perhaps that example is too cartoonish, though. Perhaps you think I'm pulling a fast one by coming up with the most morally monstrous act I can think of. And in a sense, that's exactly what I'm doing, because disproving relativism only requires showing that there's at least one case in which there is an objectively correct moral answer.

Lighting babies on fire for fun is wrong. Boom. Relativism defeated.

With the logical point made, though, it's worth noting that, sadly, there are plenty of real-world cases that put just as much pressure on relativism. Is it possible that the Third Reich was doing the right thing by waging war and exterminating minorities for the sake of creating a master race? I take it that the answer is equally obvious here as in the case of the baby-burner: no, that's not possible. Could it have been correct that slavery was ethically permissible in the pre–Civil War United States? No, that couldn't have been correct. Even though many people thought that slavery was permissible, they were wrong about this. Slavery is and was wrong.

The claim being made here is that our most confident judgments about morality rule out relativism. They rule out relativism because morality must in some cases be able to adjudicate between belief systems that disagree. For those who arrive at relativism because they are trying to be tolerant of others' beliefs, what they missed is that not all beliefs should be tolerated; not all moral claims should be respected. And while that's perhaps uncomfortable to say, it's also essential for any idea of moral progress. One of the ways in which societies morally improve is by outlawing slavery, and we make sense of this improvement by noting that society went from endorsing a false moral belief (some humans are worth less than others) to endorsing a true moral belief (all humans have equal moral value).

These are extreme cases, but if I've convinced you that lighting babies on fire for fun is objectively wrong, and that the United States became a morally better society when it freed African slaves, then we've let the camel's nose under the tent. There are some objectively true moral claims, and so now we must discover how to reason about ethics together.

When discussing ethics, many people say that they have an opinion about some moral practice. This language of opinion often gets tied up in relativistic intuitions, and so if I ask you if abortion is morally permissible, you might be tempted to respond, "Well, in my opinion it's wrong, but that doesn't mean it's wrong for other people." Or you might voice your private view but then say, "But that's just my opinion. Who am I to decide what's right for others?"

If we reject a radical, thoroughgoing relativism about morality, we must avoid this sort of language. It's tempting to think about moral judgments as similar to aesthetic judgments: we have opinions, but others will disagree, and there's not much to say in those cases. There's a saying in Turkish that I learned from my partner early on in our relationship: "Renkler ve zevkler tartışılmaz," which she translates as "Tastes and colors are inarguable." I love that saying, because it so succinctly makes clear what's special in the realm of the aesthetic, which is that arguments won't determine who is right. Whether something strikes you as funny or beautiful, whether a certain food is delicious or a color is wonderful, are perfectly good opinions, and if we disagree, that's a totally fine way to end discussion about it. Perhaps unsurprisingly, this sort of saying is not unique to Turkish. In Latin, there is a phrase, "De gustibus non disputandum est," which means "About tastes, it should not be disputed." It shouldn't be disputed, because it's inarguable.

That doesn't mean that we aren't sometimes passionate about our aesthetic views. If you, for instance, watched *The Big Lebowski* and then told me it's a terrible film because it's utterly unfunny, I would argue with you. I would bring up the clever dialogue, the perfect, deadpan performances, and generally try to convince you that it's really quite a good movie. And partly this is because film— like most art forms—has standards that we tend to accept for evaluating its excellence. So there are some features that film critics can all point to that make widely celebrated films better than widely panned films. But there's something funny (objectively?) about our disagreeing deeply about our aesthetic responses to a movie, because it's hard to believe that there is some objective fact of the matter about whether *The Big Lebowski* is funny or beautiful. It's the kind of question about which it's perfectly appropriate to have an opinion, and opinions differ. That's why we don't expect film critics to always agree, even if they tend to apply the standards of film as art more or less consistently. While a film may do something technical very well (and may be celebrated as good in that way), if it's boring, you might not like it. And that's OK.

Tastes and colors are inarguable.

In ethics, we aren't debating opinions. Ethical claims are arguable, because at least some of them are objectively true or false. While it's perfectly appropriate to conclude our debate about *The Big Lebowski* by acknowledging that it's funny for me and not for you, it's not appropriate to end a debate about the permissibility of slavery this way. Slavery is impermissible, which means that if you disagree, you're wrong. That's not me "deciding you're wrong" or "saying you can't disagree." Rather, I'm just noting that when there is a fact of the matter, believing the opposite means you're wrong. You're "allowed" to believe that 2+2=5 in some sense of "allowed" (no one will come throw you in math jail); but you're wrong, because 2+2=4.

What I'm suggesting is that "slavery is wrong" is objectively true in the same general way that "2+2=4" is. They are both facts, which means that disagreeing with them makes you wrong. And that's why doing ethics is not a squishy activity involving unsupported opinions. Ethics is the process of offering reasons and evidence in support of a claim. In ethics we don't opine; we argue, because we're aiming at the truth.

I will admit, though, that the idea of a moral fact is quite strange. Often we think of facts as empirical—as matters that can be resolved through observation and the scientific method. It's a fact that there is at least one person in my house, which I can determine by observing that I am in my house. But moral facts are not like that. We cannot directly observe them. But neither can we directly observe mathematical facts. Sure, we can see that adding two apples to two apples gives us four apples, but we cannot directly perceive the fact that 2+2=4. Moral claims, like empirical and mathematical claims, are arguable; if we disagree, there is a fact of the matter that determines who is correct.

There is much abstract philosophy written on the nature of moral facts (and mathematical ones, for that matter), and I don't pretend to resolve all of the issues in it. But I do claim that the rejection of relativism tells us that moral claims can be objectively true or false.

The empirical sciences have a methodology for discerning truth based on observation, and mathematics has a system of proofs. But moral facts are not directly observable, nor are they provable, and so ethics needs its own model for offering reasons and evidence.

CHAPTER 7

Moral Theory as Methodology?

People do not like to think. If one thinks, one must
reach conclusions. Conclusions are not always pleasant.
—Helen Keller, *Her Socialist Years*

On my very first day of college, in September of 2000, I met Dr.
John Ahrens. John is a philosopher, and he looks like one. Even
back then, he looked to me like an aging hippie, with long white
hair pulled back into a ponytail and an aggressive smoking habit.
When I walked into the classroom, he was leaning back in his
chair, with his feet propped up on the desk in front of him. John
made philosophy look very, very cool.

At the start of class, he moved to the old-school chalkboard and
began, "Philosophy is quite a simple discipline, when you get down
to it. There are really only three questions. You answer those questions, and you're done with philosophy." He was drawing vertical

lines to separate out categories, with titles at the top of three columns. "Unfortunately," he continued, "those three questions are really big. They are: One: What exists? Two: How do we know? And three"—he turned around to look at us for emphasis—"What the hell do we do about it?"

Those three big questions correspond to the fields of metaphysics, epistemology, and ethics, respectively. But ethics is about much more than just "What the hell do we do?" which is a question about action—which acts are obligatory or the subject of duty (I use these terms interchangeably), which are merely permissible, and which are prohibited. It's also the field where we investigate questions about value (the study of which is sometimes called axiology), or about what is good and bad. Rather than evaluating actions, value describes states of affairs, or the way things are. It's also the field in which we explore ascriptions of character—virtue and vice—sometimes called *aretaic judgments*, from the Greek word "arete" for excellence, or virtue, as well as the appropriateness of what are called the *reactive attitudes*, which include attitudes like blame, shame, and guilt.

As John warned: it, like the other two questions, is massive.

Not all ethical questions are of the same kind, though. In addition to those listed above, there are both more abstract and more concrete questions about right and wrong, good and bad, and so we can further divide ethics as a group of subfields arranged vertically. The questions at the top are the most abstract and theoretical, while the questions at the bottom are the most concrete, attempting to address issues that matter for the actual world of people and lived experience. That map looks like this.

Metaethics

Metaphysics and epistemology of ethics: What is the nature of morality?

How do we know?

Moral Theory

What makes an action right or wrong? What makes a state of affairs

good or bad?

Applied Ethics

What should one do in a particular instance? How should doctors behave

(medical ethics)? How should engineers behave (engineering ethics)?

How should I behave (practical ethics)?

The first two chapters of this section addressed just a couple of metaethical claims (rather quickly, it must be said), because those views about the nature of morality claimed to undermine our need to do rigorous moral reasoning of our own. My goal was only to convince you that a couple of seductive, easy-looking solutions to ethics are neither easy nor ultimately attractive. The positive project of metaethics is much more difficult, as it asks us what morality is, where it comes from, and how we access moral truths if there are any—that is, how we know what is right and wrong. This sort of abstract philosophy does not make much contact with the real world of deciding what you should do in any particular situation.

The next step in bringing ethics closer to the ground, then, is to ask what the right, the good, the virtuous, and the praiseworthy look like. If you ask a philosopher why it's wrong to light a baby on fire for fun, they are likely to respond by pointing to the suffering involved, or perhaps the cruelty of the person inflicting the suffer-ing. But if suffering and cruelty are the sorts of things that make an action wrong, then presumably they would make other actions wrong, too, right? So are all cruel actions prohibited? The drive to identify the features of an action that make it wrong has led phi-losophers to construct great, vaulting theories that attempt to pro-vide comprehensive explanations of what makes an action right or wrong. And these are the subject of moral theory.

Some philosophers believe that we need to know moral theory before we can do any practical ethics, because the way we do that work is by applying the theories in question. But I don't think that's the right way to approach moral reasoning. What we want from moral philosophy is some skill in identifying and reasoning about various moral considerations.

Imagine that you are out for a walk one day, just minding your own business, when you hear the clatter of a fast-moving vehicle behind you. Alarmed, you turn around to see an out-of-control trolley, barreling downhill. Some evil character has cut the brakes and sent the trolley out into the world in the name of an unknown, nefarious plot. You quickly look to see whether there is anyone in the path of the vehicle, and—to your horror—you see that five people have been tied to the tracks. Luckily, you are only feet away from a switch that you can throw to direct the trolley onto a spur, diverting it away from the five. Knowing you have only seconds to act, you leap toward the switch, simultaneously looking down the path of the spur to make sure it's clear. But alas, it is not. The same dastardly villain who has tied the five people to the main track has tied a single person to the spur track. (Why didn't he tie five to this track as well? Presumably to make a better thought experiment.)

If you do nothing, the train will fly by in mere seconds, smashing into the five people, killing them all. But if you throw the switch, you'll divert the train, saving the five but instead sending it to squash the poor single person tied to the spur track.

What do you do? Train coming in 5, 4, 3, 2, 1 . . .

If you're like most people, you decided that you should throw the switch. Human lives are valuable, and the more lives saved the better. So if you can save five at the cost of just one, you should do so. Most people asked in studies find this case rather easy.

But we're not done yet. Here's a slight modification to the story: it begins as before, with a runaway trolley heading toward five people tied to a track. But no switch this time. Instead, you're standing on a footbridge, directly over the trolley track. You see the trolley speeding toward the innocent victims, and it will pass below you in a matter of seconds. As you panic about what to do, you notice that there is a large man—a very large man, nearly seven feet tall and clearly a competition powerlifter—standing next to you, watching the horror unfold. Now normally, this would be unremarkable, but you happen to be a world-renowned trolleyologist, and so you happen to know that this man's mass (well over three hundred pounds) would be sufficient to stop the train, should his body find its way onto the tracks.

Of course, he's not on the tracks; he's on the footbridge above them. But he is leaning over the edge of the footbridge, watching anxiously. And—you gasp as you realize it—he is on roller skates, teetering precariously as he leans his center of gravity out over the edge. It wouldn't take much of a push at all to send him over that edge. Indeed, he might even fall over of his own accord at this rate.

So, should you give him a little push? Just help him to teeter a bit farther over the edge? If you do, he'll fall onto the tracks and be immediately crushed by the oncoming trolley, but the five people tied to the tracks will be saved. After all, human lives are valuable, and saving more is better than saving fewer. So if you can save five at the cost of just one, it seems justifiable, right? Train coming in 5, 4, 3, 2, 1 . . .

If you are like most people, despite the math looking identical to the first case, you don't think you should push the big man onto the tracks. And that is strange. Because it sure looks like the cases are morally equivalent: five people will die if you do nothing; if you act, only one person will die, and the original five will be saved. What could justify our differing intuitions?

These (in)famous cases are called *trolley problems*, and they were originally developed by the twentieth-century British philosopher Philippa Foot (though I may have taken some, ahem, creative license with the versions I presented). Though they have leaked out of the academic world and into the mainstream a bit (you may have seen them discussed on the comedy series *The Good Place*, or seen cartoon versions of a trolley problem made into a meme), they've also been extensively criticized. Many philosophers find them overly simplistic and unhelpful, and some even think that such problems turn us into bad people.

Trolley problems are popular because they look as though they tell us something about moral reasoning. When making a trolley judgment in isolation, we often seem to appeal to a generalizable principle about what is justified in the pursuit of saving lives. But the pair of cases together puts pressure on that reasoning, forcing us to examine a tension between moral commitments.

The original trolley problem, where you can throw a switch to divert the trolley onto the spur, leads most people to rely on a sort of mathematical form of moral reasoning: lives are valuable; five is more than one; and so we ought to save the five people. This is called *consequentialist reasoning*. It relies on the view that the consequences of an action can make it right or wrong, and a general belief in the idea that consequences can be added up, such that we can ask which action has the best overall consequences. Consequentialism is a fiercely intuitive view at first blush because the motivating intuitions are so compelling: we can identify some things as good or bad (for example, happiness is good, suffering and death are bad), and so we should reduce the bad things and promote the good things.

Consequentialism as a formal theory takes these intuitions one step further and holds that the consequences of an action are the only things that matter morally. Thus, the rightness or wrongness of an act is determined solely by its consequences. Further, most consequentialist theories are maximizing, and so add that the right action is the one that maximizes good outcomes and minimizes bad outcomes. Saving five lives is better than saving one, and so you should throw the switch. Easy. And now you should determine the right action in all situations in precisely the same way—identify the likely consequences, then do the thing that will make the world best overall. (Of course, we have to determine what things are good and bad, and that's what differentiates the various consequentialist theories. But we have the basic framework for now.)

The shift to the footbridge problem, though, throws a wrench into that simplistic view, because now the math is the same (five lives versus one), but most of us do not think we should push the big man onto the tracks in order to stop the trolley. But why not? What is the morally relevant feature of the case that justifies letting five people die rather than causing one to die?

In response, non-consequentialists—or, more particularly, deontologists—have offered various suggestions. In the same way that consequentialism is a sort of umbrella term, identifying all theories that make right action depend solely on consequences, deontology is the opposite umbrella camp. According to any form of deontology, particular actions can be morally required or prohibited, regardless of their consequences, because of the nature of the act. Some actions are just wrong, and that is explained not by the fact that it causes some harm but rather by the kind of act that it is. To get your deontological intuitions going, consider whether you ought to sometimes tell the truth or keep your promise, even if you might make other people happier by lying or breaking that promise. The sense that a promise ought to be kept, just because of the kind of act it is, is a deontological intuition.

Most of us have other deontological intuitions as well. We believe that we have rights, which are moral entitlements to act in certain ways, even if those ways don't produce the most good. We also believe that we are not morally required to care about everyone's well-being equally, and indeed, that we would be bad parents, friends, lovers, and spouses if we did. It is good, one might think, to spend more resources on loved ones and perhaps one's own community, even if those resources would be better used elsewhere. In short, many people believe there are more ethical considerations than simply good and bad consequences.

Returning to the trolley problem, then: from one particularly famous perspective, inspired by the nineteenth-century philosopher Immanuel Kant, one must not push the big man because to do so is to use him as a mere means to our own ends. Doing so would violate what Kant called *the categorical imperative*, which instructs us never to use anyone as a mere means but to always treat humanity (wherever it is found) as an end in itself. Kant thought that this was the supreme principle of morality, identifying the property that makes actions wrong (using another as a mere means). So we can't push the big man because to do so is to use him as a mere means to our end of saving the five on the tracks.

On a modern extension of the Kantian view, one of the ways in which we permissibly use other people depends on their having consented to being so used. I use the barista at the coffee shop to procure coffee, my students use me to gain knowledge, skills, or at least a degree, and so on, and these relationships are permissible because they are all consensual. But the big man most certainly didn't consent to being used as a trolley-stopper, and so we cannot permissibly use him.

Another common explanation of our intuitions in the footbridge problem is that the big man, like all of us, has a collection of moral rights, one of which is that one not violate his bodily integrity. By putting our hands on him, we violate his rights and so wrong him. These rights can be explained in many different ways, but if we believe that there is such a thing as a moral right, then those rights can sometimes prevent us from trying to promote the good.

Yet another explanation is that one ought not to be the cause of harm. Although the harm to the big man if you push him is less than the aggregate harm to the other five people if you don't, it is wrong to become personally responsible for some harm by doing the thing that causes it. This sort of reasoning is sometimes summarized as the view that there is a moral difference between doing and allowing harm, and so a serious moral difference between killing and letting die.

In general, then, deontologists tend to try to explain why it is wrong to kill the big man in pursuit of saving five others, and all of their methods involve appeal to non-consequentialist considerations. If some version of deontology is correct, then we are sometimes permitted or even required not to promote the best outcome. Although the consequentialist idea that we should always bring about the best world sounded so plausible on the surface, deontologists argue that there are things we shouldn't do, even when doing them would bring about the greater good.

But now the trolley problem game kicks off in earnest. Whereas it looked like the consequentialists' commitments implied that if it's right to throw the switch, then it's right to push the big man (and that seems unbelievable!), the pressure on the deontologists goes the other way: if it's wrong to push the big man, then it's wrong to throw the switch. Why? Well, because if pushing the big man uses him as a means, violates his rights, and injects one into the causal chain that causes harm, how does throwing the switch not do those things?

Most obviously, throwing the switch certainly makes one the cause of harm and death. So if you were tempted to rely on the difference between *doing* and *allowing* to distinguish between the cases, it looks like you're in trouble. Sure, pushing the big man feels like a more direct cause of a death, but throwing the switch is still the thing that makes the trolley kill the person on the switch-track. Indeed, many of those I've spoken to who are in the small group who don't think you should throw the switch reason in this way: if they don't do anything, the trolley (and the villain) kill five people, which is bad, but it's not on them; but if they throw the switch, then they kill someone. They're a killer. It's most definitely on them.

As to the Kantian and rights-based approaches, it's unclear why causing death by throwing a switch wouldn't be violating the victim's rights or similarly using them as a mere means. If each person has a right to bodily integrity, and you perform an act that sends a trolley hurtling toward them, it seems like you violated that right. (Consider if it were a missile and not a trolley—doesn't hurtling a death-object at someone seem like a violation of their rights?) And if you did it to save five other people, it seems like you used the single person as a mere means, not gaining consent but instead killing them for the sake of some other good.

It looks, that is, like both consequentialist and deontological theories have trouble making the intuitive distinction between the two trolley problems.

We can absolutely go deeper into this rabbit hole, as there is an entire cottage industry of trolley problem solutions and challenges. One major question discussed in the literature is whether a moral tool developed by Saint Thomas Aquinas (and, as a result, a central part of Catholic ethics) can help here. According to these "Thomistic" ethics, one is obligated to promote the good and avoid evil; so far sounds like consequentialism, right? Do good and avoid bad. But Thomists want to distinguish between causal roles one can play in the production of harms in cases much like the trolley problem, and so they rely on a rule called the *Doctrine of Double Effect*. The exact formulation of the rule is a bit technical, but the basic idea is that it's morally worse to intend the harms that one causes than to merely foresee that one's action will cause harm. This distinction helps in many cases (the Thomists believe), including in the case Saint Thomas originally designed it for, which is self-defense. If one does violence for the sake of self-defense—not intending the harm, but intending one's own safety—it can be permissible, even if the harm they cause would otherwise be a violation of duty. But some Thomists also think it helps with trolleys: in the original case, one intends to save the five and merely foresees the death of one as an unfortunate side effect; but in the footbridge case, one intends the death of the big man, making it a murder rather than a permissible killing.

As you can imagine, philosophers (who will argue about literally anything—I've witnessed many arguments about what makes something a sandwich) have pushed back hard on this sort of explanation. After all, the person who pushes the big man doesn't want him dead—they're also intending to save the five. What about pushing him makes it the case that his death is intended? This sort of question begins an interminable debate in the moral philosophy literature about what counts as an intention, at one point culminating in a philosopher asking whether blowing someone up with dynamite amounts to intentionally killing him, so long as the killer would prefer that the exploded person congeal back together afterward (lest you think that's a joke, I've provided receipts in the endnotes).

Some moral theorists think that our inability to justify consistent judgments in trolley cases is due to a focus on the question of obligation, duty, and moral permission in isolation. According to versions of a school of thought called *virtue theory*, the primary question we should ask is not "What should one do?" but rather "Who should one be?" Virtue theorists might argue that both consequentialists and deontologists are stymied by trolley problems because they are focusing on the wrong question. It is not the case that you ought to do some particular thing, but rather that you should be a good, virtuous person. The trolley problem (and other dilemmas) is asking the wrong question.

Of course, that might strike you as very unsatisfying, because you want to know what you should do when a runaway trolley is barreling down the tracks. This is a common criticism of virtue ethics by other moral philosophers, who argue that ethics must be action-guiding, and so should tell us what to do. In response, Rosalind Hursthouse has argued that virtue ethics gives just as much guidance as any other theory. If you want to know what to do among runaway trolleys (or in any situation), you should do what the virtuous person would characteristically do in that situation.

So. You know. Problem solved.

OK, so that sounds like frustrating advice, and if you're still unsatisfied, you're not the only one. But what virtue ethics does is allow us to acknowledge that we won't be able to generate concrete principles or give moral advice in the absence of thick context. A virtuous person has well-developed habits of character that are able to pick up on relevant moral considerations as they arise, and so seems fit for the complex reality of the world we live in. This sort of view is clearly attractive to some, as it has survived for thousands of years—it's the oldest of the theories discussed here, going back at least to Socrates, Plato, and Aristotle in Greece and to Confucius in China. But it is much less popular than consequentialism or deontology precisely because it doesn't feel like it provides the sort of concrete guidance we want from a moral theory. Modern moral philosophy is largely about the debate between consequentialists and deontologists, with the central question thus being: Is something other than the consequences of one's action relevant to the rightness or wrongness of that action?

Trolley problems, then, are often used as teaching tools by philosophers who want to frustrate their students—I mean, who want to help their students see the differences between the various broad camps of moral theory. What's less clear, though, is how seriously we should take them. Should we expect that there is a real solution to the trolley problems? Is there one theory to rule them all that will explain satisfyingly why it's permissible to throw the switch but not push the big man?

Here's where I differ from some philosophers: Given how much is written about trolley problems, it's clear that at least some philosophers really care about solving them. But I don't think there's a solution to the trolley problem, because I think it's a trick. It's a trick in a similar way to how Haidt's study question was a trick: it asks us to solve and be comfortable with some moral dilemma, but the details make it unhelpful for the messy, real world of everyday ethics. In short, these sorts of problems put pressure on moral theories, but it's not clear how much pressure they should put on actual moral judgments.

What I mean by this is that the conceptual map at the beginning of this chapter, depicting metaethics above moral theory, which is above applied ethics, invites us to make an inference about the role of moral philosophy. It invites us to suppose that we must first figure out the Truth about moral theory so that we can go on to apply it to real cases in the world. If this is the right model, then we might look to cases like the trolley problems, in which we abstract away from pesky details about real people and real life, to carefully hone our theoretical convictions. Such thought experiments allow us to isolate particular moral variables so we can try to determine what roles those variables play in moral reasoning.

But what if that's the wrong way to think about the relationship between moral theory and practical ethical reasoning? What if the real world is so much messier than trolley problems that they're not much help at all?

CHAPTER 8

Trap #3: The Right Action Is Spit Out by My Preferred Moral Theory

The relation between moral theory and applied ethics is somewhat like the relation between pure science (like physics) and engineering.

—Mark Timmons, *Moral Theory: An Introduction*

In 1972, a young philosopher named Peter Singer published a philosophy article titled "Famine, Affluence, and Morality." Singer had grown increasingly distressed by the enormity of suffering coming out of East Bengal, where a devastating cyclone had combined with what would come to be known as the Bangladesh Liberation War to wreak absolute havoc on the people there. Millions of refugees were suffering and dying from starvation, dehydration, and easily preventable diseases. In response to the unfolding tragedy, he made a few key points that—still, more than fifty years later—challenge readers and tend to cause significant discomfort to anyone willing to consider them with an open mind. His argument went like this.

The suffering of poor, displaced people is entirely preventable with resources that, in our globalized society, we can give to them; that is, we can send money and resources to victims of calamity, thereby preventing suffering and death. Although the plight of those in East Bengal as Singer was writing made this point vividly, the suffering happening then and there is not unique. Preventable suffering and death are a constant feature of our world. He next claimed that if we can prevent suffering without thereby sacrificing something of significant moral value, we are morally obligated to do so. And, as it turns out, most of us who are wealthy by global standards spend money on pure luxuries, such as dining out and going to movies. But pure luxuries do not have significant value. The result of this chain of reasoning is that each of us can prevent suffering in the world by redirecting money that we would otherwise spend on luxury goods and experiences to organizations that will use it to address extreme hardship. And since such a sacrifice would not cost us anything of significant moral value, that's what we should do.

If Singer is right, then basically all of us with disposable resources are acting badly. Every time we purchase luxury goods, we should feel the guilt and shame of knowing that we are choosing to let people die for the sake of some unnecessary consumer pleasure. Although the Bangladesh Liberation War is long past, our world is full of preventable suffering, with no end in sight. Over the years, I've thought regularly of the feeling I first got reading Singer's paper. I thought about it as the ever-more-common climate disasters destroyed homes and lives; I thought about it during times of violence and unrest, whenever there was an obvious way that I could donate to worthy causes; and I thought about it while writing this book, as Russia's army invaded Ukraine, leading to millions of refugees needing homes and resources as they fled their country. As long as suffering and need like this exist in the world, Singer's argument offers a powerful challenge to the status quo for all of us who spend our money on things that are less important than saving a life.

Many of my colleagues and students over the years have reported being moved by the argument, and it has been hugely influential not only in academia but in motivating real people to make real changes that matter. Singer has continued his work on ethics and charitable giving, and it has played an important role in the development of the effective altruism movement, which urges people to donate large portions of their resources to the most effective organizations for relieving suffering and promoting good lives.

Singer's argument—and that of many effective altruists—is rooted in a set of core, theoretical beliefs that sometimes have surprising implications but are really quite simple and easy to initially appreciate. Singer endorses the moral theory of *utilitarianism*, which is a version of consequentialism. While consequentialism tells us only that rightness and wrongness can be determined entirely by the consequences of one's actions (and typically, that we should bring about the best world), utilitarianism fills in some of the blanks by telling us what it means to bring about the "best" world. According to utilitarians like Singer, then, we can derive all of morality from just a few claims, which are:

1. Happiness is good, and unhappiness is bad (a claim about value).
2. Each of us is obligated to maximize the good and minimize the bad (a claim about what we should do in response to value).

3. Thus, each of us is obligated to act so as to maximize the total ratio of happiness to unhappiness.

4. Therefore, for each individual act (like going out to a nice meal), if the resources you would spend doing it could create more happiness elsewhere (like, say, by donating it to famine relief), then you are obligated to redirect it so as to maximize benefit.

Singer did not come up with utilitarianism as a moral philosophy—it was developed in the eighteenth century by Jeremy Bentham and has had many prominent supporters, including Bentham's protégé, John Stuart Mill, and later, Henry Sidgwick. Rather, Singer simply applied this moral philosophy in a novel context. And in fact, this is something he did (powerfully, I should note) throughout his career. In the areas of famine relief, animal ethics, abortion, euthanasia, and many more, Singer carefully showed through his writing what was implied by taking seriously an obligation to promote happiness and to minimize unhappiness, suffering, and misery.

There is a problem, though. Although happiness and unhappiness, pleasure and pain clearly matter morally, many of us believe that they are not the only things that matter morally; which is to say, many of us endorse some deontological intuitions. Utilitarianism is attractive in its simplicity: pain is bad, so we should work to make sure there is less of it; joy is good, so we should work to bring it about. But its simplicity leaves out many aspects of moral life. Don't people have moral rights? Where do they show up in the theory? (According to Bentham, the answer is "nowhere," as natural rights are "nonsense" and imprescriptible natural rights are "nonsense upon stilts.") Don't we also think that we owe some people special obligations (like parents and children, loved ones and friends), in which case we are sometimes obligated to do something even if it doesn't maximize happiness?

The important lesson here is that because Singer is applying utilitarianism, he has limited his audience. Utilitarians are easily convinced by a claim that one is obligated not to spend resources on luxuries like a nice meal rather than contributing to famine relief, because they already endorse the foundational claim that one ought always to act so as to bring about more happiness and less suffering. Each of us gets only marginal happiness from luxuries, while being relieved of starvation or being treated for an easily preventable disease provides a significant boost in overall happiness. But deontologists were never on board with that foundational claim, and so are likely to find the argument less powerful. Are you obligated to donate your money? Well, that depends. Have you promised it to anyone else? Do you have other special obligations to see to? In general, how many personal projects are we morally required to sacrifice to the insatiable maw of happiness promotion?

The point here is not to adjudicate whether you should be convinced by Singer's argument about charitable giving but rather to reveal a problem with the deductive model of applied ethics. Singer is unlikely to convince non-utilitarians of his view, and so the project has limited reach. Although I agree that most of us do less than we should to combat poverty and disease, and I believe the world would be better if more people contributed to efforts to alleviate suffering, I'm not convinced of the strict conclusion that each of us is morally required to forgo all luxuries in order to promote such efforts. And that's largely because I'm not a utilitarian. I think other things matter morally. And so as I've thought about Singer's uncomfortable conclusion, I was ultimately unconvinced by an argument that was grounded in utilitarian reasoning.

Utilitarian arguments like Singer's are certainly valuable. The argument for charitable giving has had real-world impact, and that's important. What I'm pointing to is a problem for philosophers who think we should derive all of our practical insights from theory, because theory is divisive. It breaks people into camps. And I think the tendency to push us into theoretical camps is yet another trap when it comes to doing ethics.

Let's return to where we ended the last chapter—to the relationship between moral theory and practical ethical issues. It's tempting to think that such a relationship is accurately represented by the arrows I drew on my conceptual map of various ethical issues: it's a deductive relationship. Particular ethical questions are problems of "applied ethics" because we need to determine what the correct moral theory is and then apply it to the case. As the philosopher Mark Timmons puts it, this makes the relationship between moral theory and applied ethics look like the relationship between physics and engineering: we figure out the theoretical truth and then apply it to the messy world of nonideal circumstances. Engineering is an applied science; practical ethics is applied moral theory.

If this were the right view of ethics, it would have profound (and profoundly depressing, I think) effects for real-world practical deliberation. It would mean that ethics is the sole province of philosophers, because doing practical ethics would require determining the truth about abstract moral theory before applying it. And the bad news about such an implication is that there is nothing like consensus about the correct moral theory, despite thousands of years of investigation. So practical ethics would be beholden to a project that currently employs hundreds of philosophers to continually disagree with one another about abstract details. Although this might be fun for philosophers, it doesn't sound all that helpful.

I suspect that part of the attraction of moral theories is that they promise "easy" answers to difficult ethical issues in a way analogous to how divine command and relativism did the same: once we know some general claim about ethics—the right action is what God commands, the right action is what your culture says, the right action is whatever maximizes happiness—determining the answer to a concrete practical question is just about plugging in facts. That is certainly not to say that moral theory is easy. There are some very sophisticated arguments for and against pretty much every theory ever offered. But it's the ease of being able to apply that theory that is seductive to beings like us who want to be able to justify our actions.

This attraction to the ease of application can be seen in one of the central moves in debates about moral philosophy, which is "biting the bullet." When a philosopher offers a moral theory, one of the things opponents often do is show that the theory has some uncomfortable implications when it is applied in various contexts. So utilitarians put forward their view, and all non-utilitarians attempt to show that utilitarianism has wild and unacceptable implications. We do this by offering "counterexamples," or cases in which applying utilitarianism generates conclusions that seem decidedly false. Philosophers call such a move a "reductio" after the Latin "reductio ad absurdum," which means "reduction to the absurd." If a theory implies an absurd conclusion, then we should reject the theory (or so the reasoning goes). We've actually already seen a sort of version of this with the trolley problems: utilitarianism tells us that we should save the most people in a runaway-trolley case, so opponents of utilitarianism raise the footbridge variant, in which it seems one *shouldn't* try to save the most people (since that would involve murdering the big man). But the utilitarian could always bite the bullet and accept that one *should* push the big man, however wrong or distasteful that seems.

In general, if utilitarianism is true and each of us is morally required to act in every case so as to promote the most happiness, it's not just footbridges that should worry us; there are many possible instances in which we should do something that sure seems morally horrifying, as long as we wouldn't get caught. Another famous purported counterexample to utilitarianism concerns a surgeon who recognizes that her patient (a registered organ donor) is a perfect match for five patients waiting for organ donations. The surgeon could "accidentally" make a mistake during surgery, killing her patient, but then immediately call the transplant teams and harvest the patient's organs to save the five people on the waitlist. It seems that on the utilitarian view, the surgeon is not only permitted to kill her patient, she is morally obligated to do so, since saving five lives promotes more happiness than saving one.

Utilitarians have offered all sorts of responses to this and similar cases, but crucially, one of the things they can say is, "That's right. The case is unrealistic in many ways, and a surgeon wouldn't generally have all of that information, but if they did, and the situation was real, then yes—they should save five lives rather than one." That is, they can bite the bullet and accept the unsavory implication of their view.

When a moral philosopher bites the bullet, they typically try to make the offensive conclusion go down easier in some way (by noting the lack of realism in the case, for instance), but what they're essentially saying is that the theory is more important than getting the particular case intuitively right. They're sacrificing what many people would see as a plausible judgment in one case in favor of defending their grand view. If the theory is good, then you should accept it, even when you might otherwise disagree with judgments that it generates about particular cases.

Bullet-biting reveals how attached we are to the simplicity of the applied model of ethics. If we can save the theory, then we can go on applying it to all cases. We've answered the hard ethics question in advance; all we need to do in order to evaluate different cases is plug different facts into the theory. Moral theory is seductive.

But most of the difficult ethical problems we face don't feel adequately answered by simply applying one moral theory. The simplest consequentialist theories sacrifice all our intuitions about rights and the need to respect individuals (push the big man!), but deontologists are awfully precious about getting their hands dirty, such that it can look more important not to do some bad thing oneself than to make the world a better place (don't throw that switch!). So why do moral philosophers work so hard to find the one, true, grand unifying theory? It's the seductive ease of the applied model, and we should be very suspicious of how many bullets we're willing to bite in order to save those theories that sounded so good before we started to use them in the real world.

In reflecting on Singer's famine-and-affluence argument, what I'm totally convinced of is that I'm not doing enough to relieve suffering in the world. But must I really do what he suggests? On the utilitarian view of charitable giving, the demands of morality are overwhelming. Singer's eventual conclusion in that original article is that we should each give from our resources until giving more would make us worse off than the people we're trying to help. That is, we should give to the point of "marginal utility," or when giving no longer increases overall happiness because it decreases our own more than it raises that of others. This is radical, and surprising, but straightforwardly implied by utilitarianism. For every dollar, you should do with it what would most increase happiness, and in a world with massive suffering, that won't be spending it on yourself until you, too, are suffering.

My own inclination is that there is something deeply wrong with this. Not because I don't agree with the value of helping others (I already conceded I should do more than I usually do—and if you're anything like a normal person, you probably should too) but because there are other values. I have obligations to my family, which I should meet even if using my money to reduce famine would produce more happiness. Each of us is also permitted, I think, to enjoy our own lives and promote projects that we personally care about, even if doing so is suboptimal from the perspective of happiness-promotion. And so on. All of which is just to say: most of us have many anti-utilitarian intuitions.

Now, the utilitarian can bite the bullet (as they are wont to do) and try to make their view look less unattractive than it seems, but ultimately concede that, yes, each of us is obligated to donate our resources up to the point of marginal utility. I certainly know some people who believe this. But I think the strict utilitarian view of charitable giving is false, and the primary reason it is taken seriously is because of the seductiveness of applied ethics. The much more interesting question here than whether we're all required to give up all our wealth (which moves virtually no one to action—not even most utilitarians) is how to respect the multiple values at stake. Am I permitted to forgo helping others all the time? Some of the time? How often? Are there some luxuries that are simply not worth pursuing ever? Should no one pursue artistic endeavors in a world with suffering, because all such endeavors are luxuries? What does it look like to carve out a decent life against the backdrop of so much suffering?

That last question reveals why the case of famine and affluence has dogged moral philosophers for so long: it's an early instance of the Puzzle. While utilitarianism is hard for many people to accept even in old-fashioned ethics cases (like the surgeon), a globalized world of suffering people that we can help exposes additional pressure points for the theory. Poverty, famine, and disease will not be solved by me, and I might even be skeptical that my individual contribution will matter much at all when sucked up into massive multinational charitable organizations. And yet the problem feels like one I can and should address. So I feel both obligated and impotent, and no amount of giving feels uniquely justifiable.

Unfortunately, I'm probably not going to make the discomfort with this situation entirely go away, but I am going to insist that we have to deal with it rather than pick a preferred theory (that enjoys nothing like consensus), apply it, and then simply bite the bullet when it generates counterintuitive judgments. But if we abandon applied ethics as a method for ethical reasoning, we need to come up with something else in its place.

Among fields that have engaged in formal practical ethics reasoning, biomedicine may be the most developed. Biomedical ethics, as it was originally called (focusing just on medicine and research), or bioethics (which sometimes signals broader interests, including in public health, science policy, the environment, and maybe even AI and robotics) has a large scholarship on "method." In the late twentieth century, several bioethicists engaged in a sustained discussion over years about how ethical reasoning ought to be done when it comes to practice, and that debate continues through today.

Although moral philosophers often still prefer the deductive model I described earlier, practicing bioethicists tend to reject it. Bioethics as a practice is grounded and urgent, arising from real-world cases of life and death. When physicians say that a patient is brain-dead and that further intervention is futile, but the family demands that the medical team work to keep the patient's body "alive," what is to be done? When an infectious disease outbreak hits and there are not enough ventilators, ICU space, or trained personnel to care for the sick and dying, how do we triage care? If it would help speed the development of an important vaccine, is it permissible to intentionally infect healthy volunteers with a dangerous disease to test the efficacy of novel pharmaceuticals? These questions demand answers, as failing to respond is an answer. And very few of the clinicians, scientists, policymakers, and others who are trying to answer them will have a PhD in philosophy or a preferred moral theory. That impracticality of the applied model, combined with the divisiveness I described earlier, made the era of "high moral theory" short-lived. But the methods that sprung up in its place are instructive for our present effort.

If we think of applied moral theory as deductive, applying theories to particular cases, then at least one other method of reasoning suggests itself: we could simply engage directly in the case-based reasoning itself and completely forgo moral theory. Whereas applied ethics is "top-down" reasoning, this would be more "bottom-up," beginning with cases and reasoning in the future from settled cases. In bioethics, this methodology is called casuistry and is taken to be analogous to case law, using settled judgments about difficult cases to make future judgments.

Although high moral theory wasn't well received by bioethicists, casuistry also struck many as problematic, because it didn't seem sufficiently guided by stable moral considerations. If we make particular judgments about difficult ethics cases—perhaps that physicians should not be forced to provide futile medical care—it seems reasonable to ask why that judgment is true. And there often seems to be a reason; in this case, we might think the reason is that medical resources are precious and so may not permissibly be used for a case with no expected benefit. Or that the practice of medicine should be restricted so that interventions are only ever aimed at the health of the patient, and when there is no reasonable path to patient health, further medical intervention is inappropriate. So now it looks like we're not just reasoning about particular cases, because we have identified moral principles that seem generalizable and true.

The problems with both top-down and bottom-up moral reasoning made bioethics ripe to endorse a compromise position, which was offered by Tom Beauchamp and James Childress in their seminal book, *Principles of Biomedical Ethics*. That book, now in its eighth edition and fifth decade, has become something of a bible in biomedical ethics, largely because it offers a really useful tool—what the authors call "mid-level principles" for ethical reasoning. Beauchamp and Childress endorse four particular principles (beneficence, nonmaleficence, respect for autonomy, and justice), but I'm more interested in their methodology for moral reasoning than the principles themselves. They call the four principles "mid-level" because they do not appeal to a grand unifying theory, but they still provide generalizable guidance. These are principles that enjoy broad endorsement and can be helpful for clinicians, researchers, and anyone else in reasoning through urgent moral challenges on a daily basis. The method of employing mid-level principles—and in particular, Beauchamp and Childress's four principles—came to be known as *principlism* and has had huge influence in the short history of bioethics.

High Theory
Top-down application of moral theories

Mid-level Principles
Plausible moral considerations that provide generalizable guidance

Casuistry
Case-based reasoning, from which we can establish precedent

The crucial insight from bioethics is that one need not endorse a particular complete moral theory in order to do rigorous ethical reasoning. And indeed, many on-the-ground ethicists who work on urgent matters that must be decided find the idea that we're just waiting for the grand theoretical truth to be comical. There are many real values, and those values often compete in the real world. Difficult ethical challenges are difficult precisely because there are competing values at stake; if the situation was win-win-win, we would have just made the decision, moved on, and never thought about it again. We engage in sustained debate and disagreement because there are difficult trade-offs to be made. The twentieth-century British philosopher Bernard Williams made a similar point in his criticism of utilitarianism. On his view, not everyone who disagrees with utilitarianism does so just because they think the theory generates wrong answers in various cases; rather, part of the problem is that utilitarianism seems to imply that the answers in such cases should be easy. You do the math, insert the facts about happiness, and do whatever the theory spits out. But such a view seems to get something very wrong about the phenomenology, or the experience, of ethical challenges. They're challenges because there are winners and losers and real costs to doing one thing over another.

Reconsider the trolley problems of the previous chapter. If we had correctly worked out the true moral theory, then those problems should be easy. Utilitarianism tells us to maximize the number of lives saved, so throw the switch and push the big man. A strict prohibition on doing harm with your own hands tells you to stay out of it, so let the trolley hit the five in both cases. But if these cases were even kind of real, there would be nothing easy about them—not even the original one. If you actually found yourself in that situation, all you would know is that a trolley *looks* out of control, five people *appear* to be at risk, and you could *probably* divert the trolley and kill another person. But the only thing I'm sure of in this case is that it's already a tragedy, I have no idea what I'd do, and whatever I chose, I would probably spend a fortune on therapy to help me process it. There is no right answer, lots of terrible outcomes, and the only philosophical view I have about trolley problems is that they don't tell us anything meaningful about moral theory.

So in general, my first commitment in moral methodology is to reject high moral theory and to look for something more like principlism—a collection of moral considerations that we can use to reason together. The start of that collection of tools, then, owes a lot to the classical moral theories we've investigated. Although I didn't want to take any particular value or moral claim and build a monolithic theory around it, all of the philosophies that have stood the test of time seem to offer some insight. Here's how I think of the starting moral commitments that most people will find attractive.

Consequentialism is seductive because promoting the good obviously matters. But it falls apart when it takes that plausible intuition and builds it into a great vaulting structure, insisting that it's the only thing that matters morally. Why, exactly, is it so problematic? In short, I think it's because consequentialism treats all of us like cups (stay with me for a minute).

Cups don't especially matter in themselves. Cups matter to us because they are vessels for something that matters. Various liquids matter, and so we value cups as a means of getting us those liquids. So if we take utilitarianism as our consequentialist foil: utilitarians think people are cups and the liquid is happiness. All that matters is the happiness, no matter whose cup it's in and how it's distributed, or even what you had to do to get that liquid in that distribution.

But the thing is: we're not cups. We matter, and not just as vessels. People aren't merely sacks of happiness or suffering; we're creatures with dignity, worthy of respect. There are things you absolutely must not do to people, and ways in which we ought to engage with people, apart from how doing so relates to filling them and others with happiness liquid (OK, OK, I'm abandoning the metaphor, sorry). Yes, happiness matters, as do other good consequences, but our relationships with other people dictate what sorts of things we can permissibly do in our efforts to promote those good outcomes.

As the philosopher T. M. Scanlon (whom we met earlier) says: the proper response to value is not always to promote it. Sometimes it is, sure. If joy is good, then promoting joy is the right response. But people have value, and the right response to that value is different. Just because people are valuable, that doesn't mean we should all make as many babies as possible ("Look! I'm promoting value!"). The proper response to the dignity of other people is to respect them, and Kant—as well as other deontologists—has a lot to teach us about how to do that. Scanlon himself offers that we respect others by acting in ways that we could justify to them—a Kantian idea and one that has inspired much of my own thinking.

Now, character clearly matters, too, as we utilize character language all the time in our moral lives. When we call someone an "asshole," we are making a moral judgment about them, as we are when we talk about jerks or selfish people, as well as when we praise someone we admire as kind or generous. Indeed, many of us are more likely to describe virtues or vices in our everyday lives than we are to talk about duties or obligations. However, it does seem to get left out of many modern moral discussions, and the reason typically given goes back to one of my comments in the last chapter, which is just that it doesn't seem as helpful. We are often trying to figure out what we should do, and so advice about which virtues to have doesn't quite answer the question.

A practical moral methodology, then, appeals to multiple, competing considerations whenever they are applicable. You would be forgiven for wanting a precise set of rules for how to do that, but alas, I won't be giving you one. There is no fully optimized algorithm. The principlism of Beauchamp and Childress attempts to tell us how to apply different principles, telling us to engage in practices they call *weighting and balancing* any principles that are in tension with one another, and *specifying* the content of principles to make it clear how they apply in a concrete instance. Despite the ink spilled trying to spell out that process, however, I'm not sure it amounts to much more than this, perhaps unsatisfying, guidance: reason together from shared moral commitments.

We're trying to find a shared starting point, and then using rules of good, rigorous reasoning to move from there to more specific conclusions. The method of ethics is, essentially, moral argument. Not argument in the sense of yelling at each other but argument in the logical sense of offering claims that justify a conclusion. We're going to want to respect the intuitions behind all of the great moral theories, and we'll do that by appealing to competing moral considerations in a way similar to the principlism of biomedical ethics. This is the beginning of a moral methodology.

Is it enough to solve the Puzzle? Is there a duty not to joyguzzle?

PART III

SOLVING THE PUZZLE

CHAPTER 9

Trying to Solve the Puzzle
with Old Tools

One way to confirm the truth of my moral intuitions
would be to derive them from a general moral principle . . .
The problem is, which principle?
— Walter Sinnott-Armstrong, "It's Not My Fault:
Global Warming and Individual Obligation"

When I think about the Puzzle, I tend to think about it in a fairly loose way. It's about feeling powerless to have any effect on large, structural issues. It's about how small each one of our private actions is, and how large and complex our global, interconnected society is. So many of our choices lead us to do the sort of thing that, when lots of other people do it, causes massive harm. But I'm just one person, and my tiny contribution to these problems doesn't seem to matter.

In the context of climate change, scientists think the Earth will warm an average of 2 degrees Celsius when humans have emitted a total of a few trillion tons of CO_2, and the result will be catastrophic—harming, displacing, and killing millions of people.

But even my most luxurious private actions, like flying across the Atlantic to see my family, contribute an infinitesimal fraction to the problem, such that choosing not to do so will almost certainly not have any meaningful consequence. And the majority of my actions aren't anywhere close to being as impactful as flying. Everyday decisions tend to be more at the scale of passing up meat in favor of plant-based food options, turning off lights when I leave a room, or driving less. Faced with the discrepancy between the global scale of the problem, then, and the fractional changes I can routinely make, I feel powerless.

Armed with our moral philosophical tools, we can be a bit more precise about the structure of the Puzzle. What makes the issue of individual ethics amidst big collective problems puzzling is the combination of two claims: (1) it seems to many of us like we have obligations not to make small contributions to massive collective harms; but (2) it's mysterious how to justify that obligation, since your individual action is so small as to be practically meaningless. Solving the Puzzle will require either denying that we in fact have such obligations (and ideally explaining why it seems that we do) or providing an explanation of that obligation—that is, rejecting either (1) or (2). We must either embrace the depressing conclusion (Joyguzzle away! You don't matter!) or we must develop a novel account of the supposed obligation (Win the Nobel Prize for Philosophy!).

Given that the depressing conclusion is, well, depressing, many philosophers have attempted to avoid it. The most straightforward way to do that would be to take the tools of classic moral philosophy from the previous section and show how they justify an obligation not to joyguzzle. Alas, this is not, it turns out, an easy task.

(Also, just kidding about the Nobel for Philosophy; that's not a thing.)

The aspect of climate change that makes it feel morally urgent is that it will cause harm, which is to say that the consequences of climate change will be bad. So using the handy language of the previous section, we should expect the moral principle that prohibits contributing to climate change to be consequentialist. If joyguzzling is wrong, then it's because of the harm it causes. We are obligated not to cause harm, so we shouldn't joyguzzle.

The problem, as we've now seen, is that joyguzzling doesn't directly cause harm. It contributes a tiny amount of GHG into the atmosphere, where a phenomenally complex, global carbon cycle of unimaginable scale will incorporate it, and if billions of other people also emit GHG at a certain level, then together these emissions will cause harmful disruptions. So it doesn't look like my individual joyguzzling is wrong due to any harm it causes.

Maybe joyguzzling is wrong because it makes the harms of climate change a little bit worse? Even if my action doesn't directly cause harm, maybe it makes some heat wave a very tiny bit warmer, or makes some storm a very tiny bit stronger? This sort of reasoning relies on the bathtub metaphor used earlier, in which each tiny contribution makes climate change a tiny bit worse in a direct, linear way. But climate change isn't a direct, linear process, and it operates on a scale that's likely insensitive to such small contributions.

In general, the challenge to individual obligation explored in this book is precisely that some outcomes are so large and so complex that my individual contribution to them isn't meaningful. In which case, there is a particular challenge to explaining the wrongness of our actions in terms of the harm that they cause, since that causal link appears to be broken.

There are certainly complicated attempts to develop such an explanation, though. Recall that some moral philosophers are consequentialists, which means that if it's wrong to joyguzzle, there must be an explanation of that wrongness in terms of the harms it causes. One attempt to provide a consequentialist explanation comes from the philosopher John Nolt, who argues that everyone who emits GHG should be assigned a proportion of responsibility for the harms of climate change, based on how much they emitted. In other words: we should add up all the human suffering that results from a certain amount of emissions, then divide up responsibility for that suffering based on each person's share of emissions. If we do this, Nolt estimates, we should conclude that the average American is morally responsible for the suffering or death of one to two people over their emitting lifetime. If such a calculation makes moral sense, then it could be really important in the fight to motivate individuals to change their ways. Killing a couple of people is a big deal! If our lifestyles are morally equivalent to doing that, then we all have some serious soul-searching to do.

Another way to try to explain how my individual action is related to collective harms is by invoking the idea of a "threshold." Although it's likely true that my individual emissions don't directly cause any harmful climate event, perhaps there are thresholds that trigger certain harms when passed. So when a certain amount of GHG is emitted, the temperature changes to such a degree that various harmful outcomes occur. If this is how the climate works, then perhaps the moral math is a bit easier: at each threshold, we account for the harms caused and assign individual activities a probability of causing emissions to cross that threshold. Then, as long as the math works out, emitting actions are wrong because the probability of causing serious harm is too high.

Borrowing from the late philosopher Derek Parfit, I call any such attempt to explain the wrongness of actions like joyguzzling "moral mathematics" arguments. In general, the idea behind all of them (and there are more than I mentioned here) is that individual contributions to massive collective harms are wrong because your action gets mathematically assigned some percentage of the bad outcome. This is supposed to salvage the idea that such actions are wrong because they caused something bad.

I don't think they do, though. The idea behind moral mathematics solutions is supposed to be that causing some minute fraction of harm, some likelihood of harm, or even some "statistical harm" is morally equivalent to causing some amount of actual harm. But they just don't seem equivalent. These solutions are clever, for sure, as numbers can be persuasive—assigning a numerical degree of responsibility, or "proving" that one's act is equivalent to something clearly morally wrong, feels precise in a way that's satisfying. But the worry is that such precision is misleading. It's hard to understand why we should accept that moral responsibility works out mathematically in the way that such views suggest.

We tend to believe that responsibility tracks the harm we *actually* cause; if I punch you in the nose, the fact that I'm the one who hurt you seems like a good explanation for why what I did was wrong, why I should be blamed and punished, and why I owe you an apology. Or in some cases, responsibility tracks the *risk* of causing harm, like in the case of drunk driving; though you may drive intoxicated without hurting or killing anyone, the risk of doing so was far too high to be acceptable, which is why it's wrong. But nothing like those stories exists for my joyguzzling. My responsibility here is mathematically made up, disconnected from actual harm caused. Every time I punch someone, I cause harm. And if I were to drive drunk enough times, I would be very likely to eventually cause harm because the risk is so high. But small emitting activities aren't like either case, since such actions neither directly cause harm nor do they have a high risk of directly causing harm if done enough. Global emissions together lead to an environment with more violent disruptions, and those disruptions cause harm. The causal link between my actions and that outcome is just too tenuous to support the weight of obligation.

The only reason I can see for accepting the sort of argument above is if I were already committed to an abstract theoretical view on which an action's wrongness must be explained by its consequences. And indeed, the philosophers who try to make such arguments work tend to be committed consequentialists. But having some incredibly tiny statistical or probabilistic relationship (that I often won't have enough information to determine) to some larger outcome isn't what seems wrong about going for a drive, eating meat, passing up the recycling bin, or failing to turn off the lights when I leave the room.

I'm not sure what I'm thinking about when I get hit with a pang of guilt for some minor environmental infraction, but it's not complex calculations. If the answer were to be found in the moral math, then we should expect that most of us would never know whether our individual contributions were ethically prohibited, because the math required involves such massive calculations based on information we don't have. But we should also expect that *were* we to have such information, the answer would be clear. But neither of those features match with my moral experience. I think I often know whether or not my individual action is problematic, even without having the information necessary to do the math. And yet when I do have access to more information that should make the math easier, that doesn't seem to make the moral situation clearer. My minor environmental infractions seem "problematic" or "lazy" regardless of the math, and they don't seem to come into focus as obviously wrong or permissible regardless of how much information I have. If I learned that some individual action had a 0.0000000001 percent instead of a 0.000000001 percent chance of increasing the intensity of a future storm by 0.01 percent, I would not take that to have clarified my moral situation. And that leads me to think that the consequentialists are looking for the wrong sort of solution.

The philosopher Julia Nefsky argues that attempts to explain the wrongness of joyguzzling can take one of two tacks. On the first, they can try to show that joyguzzling really does make a difference, whereas on the second, they can argue that an act can be wrong without making a difference. The aforementioned moral-mathematics explanations are all instances of the first strategy, but I've suggested that they're not particularly successful. That means we should ask whether joyguzzling might be wrong even if we accept that it makes no difference. In the terminology of Part II of the book: we need to look for non-consequentialist, or deontological, principles that rule out joyguzzling and the like.

At first glance, this sounds like a promising strategy. If we simply reject consequentialism, it doesn't matter that our individual actions don't, by themselves, cause significant harm. Indeed, when I explain the Puzzle to other philosophers who haven't thought about it before, they are often unbothered by it at first, responding something like, "Well, it doesn't seem super problematic that I don't make a difference, because I'm not a consequentialist." They assume that it will be easier to explain the wrongness of our actions in deontological terms. But there is reason to think that many of the classic non-consequentialist strategies will similarly struggle.

Consider one of the most prominent moral principles from deontological traditions: respect for persons. A requirement to respect others can be spelled out in many different ways, from a Kantian rule to treat all people as ends in themselves and not merely as means, to a more contractualist principle to act only in ways that are justifiable to everyone. If contributing to massive harms is disrespectful in some way, then we would have a solid, pedigreed explanation for its wrongness.

The problem is that it's totally unclear why such minute contributions to massive, structural harms would be disrespectful. Driving to the gym isn't necessary, but it does make my life better, so I do it; but each drive contributes infinitesimally to climate change. So is taking such a drive disrespectful? If it was causing harm to others, then the disrespect may be explainable in terms of not taking that harm seriously enough, or seeing other people as not mattering sufficiently to count in favor of restricting my actions. But our problem is precisely that driving doesn't seem to cause climate harms. So if it's disrespectful to drive more than strictly necessary, there would need to be an explanation that didn't rely on the consequences of the act.

Many acts do seem disrespectful, even if they don't cause harm.

Lying or breaking a promise is often taken to be wrong, even if doing so wouldn't hurt anyone, because they are actions that fail to treat others as whole persons, worthy of consideration. The Kantian sort of view here is that in lying or promise-breaking, I use people as tools for my own end. But making small contributions to massive harms doesn't seem like it's using people. My driving doesn't seem to be about other people at all.

One way to think about a non-consequentialist requirement to act or not act in a certain way is that it explains the rightness or wrongness of the act by reference to the nature of the act itself. Many moral philosophers reject consequentialism because it can justify any action, no matter how heinous that action is. Torture can be justified by preventing more torture; lying can be justified by preventing others from lying. And so the deontological intuition is that some actions are just wrong in themselves.

The Puzzle is challenging precisely because it's about actions that don't seem wrong in themselves. Indeed—our individual contributions to massive harms are often actions that are good in themselves. Traveling to see my family is good; the moral question is only raised because traveling emits greenhouse gases. Socializing and being in close contact with others is good; the moral question only arises during a pandemic because close contact spreads the virus. Many of these actions, then, simply don't seem disrespectful, because they don't meaningfully affect others and they are not bad (and may even be good) in themselves.

Another non-consequentialist concept some think might be helpful in solving the Puzzle is fairness. Intuitively, it seems plausible that my lifestyle is unfair in some way. As someone who is wealthy by global standards (simply by virtue of being a middle-class American), I eat and travel too much, I live in too big a house, and I buy too many things (to cite merely a small sampling of my environmental sins). So one might argue that, given the global limit on emissions that would be required in order to avoid catastrophic climate change, my overconsumption and flagrant emissions are unfair. I'm taking too much of a small pie.

Although this has some intuitive pull, it, too, is difficult to pay off as an explanation. In order to make sense of the idea that I'm taking more than my fair share, we would need to know what fair shares look like. And of course, we could come up with a scheme. But with reference to climate change, we haven't. I don't have a personal carbon budget that has been allocated; there are no publicized rules or norms; and so there is no group effort ongoing for me to join. What that means is that if I cut my emissions, I'm doing something that in isolation will have no meaningful effect on the outcome, because I can't expect anyone else to. I'm just making my own life harder while others continue to emit away. This starts to sound less like doing what's fair and more like being a sucker.

If this doesn't quite grab you yet, consider the analogy of paying your taxes. Each of us is obligated to pay our taxes because they are part of a standardized, publicized, and enforced scheme under which individuals contribute to the common good. Sure, my piddly contribution wouldn't be missed if I could figure out how to dodge paying it; but it would still be wrong because it's unfair. When everyone else is following the rules for the public good, failing to do so is failing to do one's part.

But suppose (as is true) that I think my own taxes should be higher. I think this because I think the American tax system is insufficiently progressive, and someone who makes as much as I do should carry more of the public burden. Does that mean I should cut a check to the IRS for more than I owe?

I don't think it does. Paying more than my share of taxes when there is no rule for doing so is an unhelpful strategy for solving the problems of America's tax code. My extra money won't make a difference, and I'm not part of an intentional collective effort to promote the good. It seems clear that fairness doesn't obligate me to pay more taxes than I owe, even though it would obligate me to pay higher taxes if the tax law changed to become more progressive.

If climate policy required certain restrictions, then I would be obligated by fairness to do my part. In fact, we may even be able to generate duties of fairness by just establishing widespread norms (informal rules that do not carry the force of policy or law). In my neighborhood, we have a community association that requests annual dues in order to keep up public lands and pay for community projects. The association does not have any legal authority, and my family could opt out. But most of the community does pay in, which is how we have the funds to pay for important projects. Given this reality, I think we have an obligation to pay our dues as well, since the norm of paying allocates to us a fair share.

The problem with climate change, then, is that there are neither formal rules nor informal norms that are widely held and endorsed. As a result, I don't think it's unfair for the average person to engage in typical, individual actions, even when they contribute to climate change (as most of them do). Fairness can and should get ahold of us, eventually, if and when there is sufficient energy behind a climate movement that individuals have some guidance on what is reasonably expected of them. So we desperately need climate policy and new norms to help clarify our individual ethics, but that doesn't help us in the meantime.

A final ethical tool from the last section that we might expect to do some work to solve the Puzzle is that of virtue and vice. And while there is an interesting and compelling literature on the role that character should play in climate ethics in particular, the same basic problem that arose for the other non-consequentialist arguments seems to apply here.

Virtue and vice would solve the Puzzle if it could explain why I have a duty not to joyguzzle, or to eat vegan and what have you. Now, on the one hand, we might think that the move to character is quite promising, since we do seem to think that people who reduce their emission and consumption activities are good in some way—they are temperate, mindful, moderate in their use of resources, and so on. But to get a duty, we have to go beyond describing someone's character. Recall that philosopher Rosalind Hursthouse argues that virtue ethics does tell us about duties, as one is obligated to do what the virtuous person would do and avoid what the vicious person would do.

The problem here, as it was in the case of respect, is that it's totally unclear why the virtuous person would always or characteristically avoid tiny contributions to massive problems. If the actions caused harm, then sure, we'd know why the virtuous person would avoid them; and if the actions were bad in themselves, then yeah, we'd know. If they were unfair? Yup, that'd do it. But what about these otherwise innocuous or even good actions makes them the sort of thing that virtuous people would avoid? This question is especially biting, since virtuous people (it would seem) often act contrary to our environmental sensibilities. Would the virtuous person fly internationally regularly to see their family? That seems like the kind, loyal, loving thing to do. It seems like it's what a good parent or sibling or child would do. But that means that "doing what the virtuous person would do" will not always rule out making small contributions to climate change.

That is the philosophically rigorous version of the Puzzle: we seem to have a duty not to contribute to massive harms, but the classical tools of moral philosophy don't provide an explanation of that duty. What makes climate change bad is that it causes harm. And that only happens through massive collective action. If my otherwise permissible action doesn't cause any of that harm, it's difficult to know why we are duty bound to refrain from it.

To solve the Puzzle as now defined, we must begin by recognizing that many of us have a rather small set of ethical tools and language.

In particular: we are obsessed with obligation and duty.

Much of the classical debate in moral philosophy concerns explaining the rightness or wrongness of an action—that is, whether it's obligatory, permissible, or prohibited. This is certainly an important category, and much of our moral lives revolve around it, as we blame and often punish those who violate their obligation. Duty and obligation, then, can be used as a kind of bludgeon. If we can show that someone else did wrong or failed to do their duty, this puts us in a position of moral authority over them. We reasonably want to know what is wrong, not only because we should avoid it but because we want to hold accountable those who don't.

Although the language of virtue and vice is part of what I'm calling traditional moral philosophy, the desire to use even the concept of character to explain right action (that which the virtuous person would characteristically perform) reveals the extent to which we try to use our tools to determine right and wrong, duty and prohibition. These are the cleanest of lines in our moral reality, and it's very satisfying to draw them.

The Puzzle is uncomfortable for so many of us because it sure seems like there is something wrong with contributing, even a little, to some massive harm, or failing to do one's minute part in a larger moral project. And it would seem very strange if *no one* should do the things that we should all do to solve a collective problem. So we reach for the tools we have ready to hand to try to explain the wrongness of our contribution, or why we must do a certain thing.

But look closely at the way I just put the discomfort raised by the Puzzle: we think there is "something wrong" with our tiny contributions, and that we "should act" in certain ways. This is fairly modest language, and yet in trying to explain precisely what that means, philosophers have often reached for the blunt tools that prove far more than is needed: they try to demonstrate that making small contributions is wrong because it's a violation of duty, and so we are morally required to act in certain ways.

This recognition of the sort of over-powered nature of our classical moral tools is the beginning of solving (or better: dissolving) the Puzzle. Too many of us—in our everyday thoughts, discussions, and social-media arguments—are obsessed with duty, and that leads us to think about our moral lives with an impoverished ethical vocabulary. Not every action that is bad in some way is prohibited; and the fact that some act would be good, or even that you should do it, doesn't mean that it's required.

Moral reality is richer and more complicated than all of that. Exploring that richness will, I think, help us to determine how to live in the world we now inhabit.

CHAPTER 10

The Wonderful Variety of Ethical Concepts

There are more things in heaven and earth, Horatio, than are dreamt of in your philosophy.
—William Shakespeare, *The Tragedy of Hamlet, Prince of Denmark*

The abortion debate occupies a strange place in moral discussion. In many parts of the world, the permissibility of abortion is the subject of vehement disagreement. The topic can be politically polarizing, and those who hold different views may see the other side as not just wrong but egregiously wrong—sometimes monstrously wrong. Laws around abortion are similarly contentious, as can be seen in the recent decision by the U.S. Supreme Court to roll back decades of precedent in its judgment in *Dobbs v. Jackson Women's Health Organization*, taking away the constitutionally protected right to an abortion and leaving legislation up to individual states.

Despite this deep disagreement, which you might expect to indicate the complexity of the topic, many view the case of abortion ethics as fairly straightforward. The argument for whatever their position is seems simple and obviously correct. This can be seen in some of the rhetoric around abortion. On the anti-abortion side, you hear things like "Abortion is murder." If that's true, it's a bit of a discussion-ender. Murder is wrong, so abortion is wrong. Case closed. But on the opposite side, we often hear something like "My body, my choice." And this, too, seems pretty straightforward. If you have the right to do what you choose with your body, and you're the one gestating a fetus, then you may choose to stop gestating. Bodily autonomy straightforwardly implies the permissibility of abortion.

The problem with both of these arguments—and I beg for your patience and charity when reading the next few sentences—is that both are obviously problematic. I don't mean this to be contentious, and I certainly don't mean this to imply that we already know which position is correct. My point is a strictly logical one: neither argument, condensed into a slogan, is actually convincing at all under closer scrutiny. And that's because the case of abortion truly is complex. It is, as we should expect from the radical and deep disagreement, genuinely hard to think clearly about whether it's ever permissible to end an early human life.

Let's start with "My body, my choice." I think there is something deeply correct about this, and I expect most people agree. In general, we have broad permission to do what we wish with our bodies. The problem, however, is that we don't have *universal* permission to do what we want with our bodies. And this, too, should be uncontroversial. The fact that I may use my body as I see fit does not imply I may use it to injure or kill another person. As J. S. Mill famously articulated in his "harm principle": your rights end when exercising them would cause harm to another. Or, more pithily:

"The right to swing my fist ends where your nose begins." Thus, the argument from the pregnant person's* right to bodily autonomy only goes so far. To be convincing, it needs to say something about the embryo or fetus being killed. In particular, it needs to show that, while a right to use your body as you wish doesn't grant you permission to kill me, it does grant you permission to terminate a pregnancy.

What this point makes clear is that a central question of abortion ethics is the question of moral status: At what point does early human life become morally like us, thereby exerting similar moral claims? Or, another way to put the point: When does the fetus become a person, with full moral standing, including a right to life? "Person," here, is a moral term, distinct from the biological category of human (or, as the philosopher John Locke put it, person is a "forensic term, appropriating actions and their merit; and so belongs only to intelligent agents, capable of a law, and happiness, and misery"). Whereas "human" denotes simply whether one is of a certain species, "person" denotes whether one has the moral standing we tend to accord to humans. It is an open—and much debated—moral question as to whether all and only humans are persons.

---

*My choice of language here might catch enough people out that it's worth a somewhat lengthy aside to note that I'll be using the terms "pregnant person" and "gestating person" rather than "woman," since not all people who gestate and give birth are women. Trans men and nonbinary individuals can and do gestate and give birth. However, this should not render invisible the fact that it tends (now and historically) to be women who gestate, which means that a feminist analysis is relevant, since it is easier to silence populations with unequal access to power. In a sexist society, the fact that women largely have done and continue to do the gestational work and endure the risks of childbirth almost certainly makes it easier to dismiss those hardships and risks. Given that gender minorities are also marginalized groups in society, this point continues to hold, even if "pregnant people" rather than "pregnant women" is the relevant group whose rights are being debated when it comes to abortion.

Those who endorse abortion rights hold a variety of positions on moral status. One prominent example is the view that personhood—and with it, a right to life—comes into existence with certain properties, such as the ability to feel pleasure and pain (often called sentience) or some capacity for sophisticated cognition. On such views, a fetus before the development of these properties has no moral status at all, while after developing them it has the same moral status as a child or adult. The achievement of some important property thus is a threshold of sorts, the passing of which entails that the fetus becomes a person. On another view, the property of personhood comes into existence gradually, over time (a view called *gradualism*), such that an early fetus can be partially a person, and thus deserving of some moral consideration, while a later fetus can be very nearly a full person, worthy of similar consideration to a newborn.

Regardless of the details, those who support abortion rights tend to be dismissive of the idea that the earliest human life is anything like the life of a person. After all, in the earliest days of human life, a zygote (fertilized egg) and blastocyst (the hollow ball of cells that implants in the uterus) are incredibly tiny, lacking any of the complexity or fascinating properties of fully developed humans. As an embryo (starting around day ten of development), it begins to develop organ systems and bodily structures, but it will not even be a fetus until about two months later; early embryos can even be frozen and kept for later implanting using in vitro fertilization, and there are millions of such leftover embryos in freezers all around the world. Something as morally uninteresting as a small clump of cells (that we are willing to freeze—sometimes indefinitely) is not a baby, and so abortion of at least early embryos is not the murder of a child. Proponents of abortion rights thus hold that early human life does not have the moral status that undergirds a right to life, and so killing it is not prohibited.

On the other side, the idea that abortion is murder begins with the opposite view—that the early human life being gestated is precisely like me, or you, or anyone else, and so choosing to kill it is just like killing one of us. Which is to say: it's murder.

There is variation among views in terms of just how early human life achieves moral status—perhaps from the moment of conception, or at some other very early marker, such as the point at which "twinning" is no longer possible (that is, when it is no longer possible that the one embryo will develop into two children). On any such view, all or nearly all early human life is the life of a human person, and so all fetuses have a right to life. This could be because humans at an early stage of development, like adult humans, have a soul; or it could be for more secular reasons, such as the idea that embryos and fetuses are "potential people," or that even at that early stage, fetuses have a "future like ours" and so are relevantly similar to us adult humans. Regardless of the explanation, what holds this camp together is the claim that the moral status of the fetus gives it a right to life, and the intentional violation of a right to life constitutes murder, and so is wrong. Thus, all abortions, which involve killing the fetus and thus violating its right to life, are wrong.

When the abortion debate is framed this way, it looks as though the determinative issue for the ethics of abortion is the question of moral status. If at least some early human life does not have full moral status, then the rights of the pregnant person may well entitle them to terminate a pregnancy. Whereas if all early human life has full moral status, then all abortions are impermissible. This is the classic setup for investigating the ethics of abortion. But it's still missing something.

When the anti-abortion side claims that abortion is murder, there is another premise being assumed in addition to their claim that all early human life is the life of a person. They also claim that it's wrong to kill a person. This premise often goes unnoticed because it seems so innocuous. Of course it's wrong to kill a person! Right?

The twentieth-century American philosopher Judith Jarvis Thomson made a permanent impression on the abortion debate by demonstrating that this premise, too, is contentious. First, we should simply note that not all instances of killing a person are murders. There are killings in self-defense, killings in a just war, and depending on your view of capital punishment, you may be committed to the view that state-sponsored killing as punishment is a non-murder killing. Such examples show that there is logical space between the concepts of intentional killing and murder, but Thomson focuses on a more relevant case: that of withdrawing life-giving care.

At a general level, Thomson notes that we are not typically entitled to all life-giving assistance. In a delightfully dated example, she posits that if she were deathly ill, and nothing would save her except the cool touch of Henry Fonda's hand on her feverish brow, he would not thereby be morally required to come to her rescue. It would be awfully nice of Mr. Fonda to fly all the way out to see her from Hollywood (where he was presumably working moviemaking magic), but she's not entitled to that act of assistance from him.

This claim is important because gestating a fetus is an act of giving it assistance. So on Thomson's view, the question of abortion is not whether one is entitled to come in from nowhere and kill the fetus, as if it were safe from harm if you do nothing. Rather, the status quo is that the gestating person effortfully promotes the life of the fetus, assisting its development. The question of abortion, then, is whether it's permissible to withdraw that assistance.

She illustrates her point with a now (in)famous example of a violinist. Imagine, Thomson asks us, that you wake up one day to find yourself hooked up to a famous violinist. You find out that this violinist was close to death and could be saved only by being surgically attached to the right person, whose biological match would allow them to use your organs to heal over the course of nine months. So the Society of Music Lovers kidnapped you and plugged you into this person, and you are now acting as a life-support system. If you were to disconnect from them, the violinist would die within minutes. Are you therefore obligated to remain hooked up to this stranger for nine months?

Thomson thinks it obvious that the answer is no, and that you may absolutely turn around and unplug yourself, declining to give this person life-giving assistance. Sure, she admits, the violinist is a person and therefore has a right to life; but a right to life does not imply a right to all life-giving assistance. You may thus permissibly withdraw assistance, even though doing so will result in their death.

Although Thomson doesn't situate her point in this way, she is relying here on what has, in recent decades, become the common medical-ethics dictum that there is a moral distinction between withdrawing life-giving assistance and actively killing someone, such as in end-of-life cases. When a patient is on life support and will never recover, it can be permissible to turn off life support (colloquially: pull the plug), knowing that this will result in their death, even when it would not be permissible to kill the patient by active means—by, say, giving them too much morphine. In both the end-of-life case and the beginning-of-life case, there is a crucial moral difference between killing and withdrawing support, thereby allowing the patient to die. And this is what Thomson thinks the anti-abortion side missed: even if abortion causes the death of a person, that simply doesn't mean it's murder.

Thomson's article is now decades old, and there is a rich literature debating what she got right and wrong. My interest, though, is not in adjudicating these details. Rather, I began this section by noting that both sides of the abortion debate often treat the issue as though it's simple and straightforward, and I think it's clear that it's not. Neither argument works unless we have a sophisticated account of moral status. And then, even if we do, we must determine which instances of causing a person's death are morally wrong.

In my own thinking about abortion, I took Thomson to be the height of nuance until I met Maggie Little. Maggie is a philosopher and ethicist at Georgetown University who, when I was there and worked as her research assistant, was also the director of the Kennedy Institute of Ethics. She is well known and well respected for work on many philosophical issues, but two of her main areas of expertise are abortion and what we can think of as the complexities of moral concepts. As it turns out, she thinks these two topics are related, and so what we need in order to investigate abortion is a much richer set of conceptual tools.

When I first heard Maggie talk about abortion, the thing she did that changed my entire view of the debate was to largely ignore the traditional framing. She did not call the issue "abortion ethics." Instead, she suggested that the question of *gestation ethics* is whether or not a pregnant person has a positive duty to create a child out of an embryo and thereby to become a biological parent. What we need to explore, she argued, is the ethics of creation. Is it ever obligatory to create a person? Is it ever good to do so? If so, when? By asking what is morally required of a pregnant person, we are invited to consider more than just the moral status of the fetus (though that will certainly be important); we are invited to consider the nature of gestation, and what it is appropriate for morality to say about it.

Maggie's framing turned the abortion debate on its head, and it has influenced far more than my views on reproductive ethics. It has influenced my entire approach to moral philosophy.

Imagine that you are tending bar in a war-torn country, and you hear a group of rowdy soldiers talking about their upcoming mission. Although they at first seem in good spirits, responding to queries with humor, the gallows are definitely present in the background. As you serve them drinks throughout the evening, you come to realize that the soldiers are steeling themselves for what they all believe will be a suicide mission the next day. They are committed to doing it—because they're good soldiers, and their cause is a just one—but they also anticipate that it will cost their lives. Now, let us assume a few things about this case: one of the soldiers, in particular, has expressed sexual interest in you; you (correctly) believe that having sex with them would provide some comfort on this—what may be their last—night on Earth; but also, that this sort of comfort can sometimes actually improve a soldier's odds of survival, by boosting morale (you must be really good at this!). Finally, assume that casual sex is something you enjoy, and that you have engaged in this sort of last-night-celebration-morale-boosting sex before, and generally think it's a kind and generous thing to do.

Given all of these features of the case, is it your duty to have sex with the soldier? Are you obligated, and therefore morally required to give of yourself in this very intimate way?

Maggie's judgment in a similar case is that the answer is no, and for a very simple reason: sex is not typically obligatory. It is not the subject of a duty. It may be good, beneficial, helpful, and kind, but even if you could save a life with sex, you are typically not obligated to do it.

To understand why this might be true, we need to think for a minute about these concepts of "duty" and "obligation," which I'm here using interchangeably (so to have a duty to do something is to be obligated to do it). When an action is obligatory, several other things tend to be true. Although philosophers disagree about whether all of the following features always attach to duties, they are a collection of the attempts made to understand the special status of our strictest moral concepts.

According to what is called the *correlativity thesis*, duties tend to have correlated rights-holders; therefore, if I have a duty to have sex with someone, that person has a right to sex with me. A right is a form of entitlement, and so that person is entitled to sex with me. Further, because of this entitlement, we might think that others in the moral community have the standing to demand that I fulfill my duty; or, as the philosopher Stephen Darwall famously put the point: members of the moral community have the authority to demand my compliance, and I am therefore accountable to everyone for so complying. Although that doesn't mean that anyone is permitted to force me to do my duty (just because I have a general obligation not to lie doesn't mean anyone else is allowed to administer truth serum to me, for instance), they are entitled to demand that I give them what is owed. Finally, when someone violates their duty, the moral community is entitled to hold them accountable by, for instance, calling them out, demanding an apology or reparations, blaming and/or shaming them, or even punishing them if they have the power to do it (perhaps informally by altering their relationship with the wrongdoer, or more formally by excluding them from some community goods). While there might be interesting cases where wrong actions should nevertheless not be blamed or punished, for the most part, John Stuart Mill seemed to be onto something when he wrote, "We do not call anything wrong, unless we mean to imply that a person ought to be punished in some way or other for doing it; if not by law, by the opinion of his fellow creatures; if not by opinion, by the reproaches of his own conscience."

All of these features of duty and obligation make up the importance and the force of moral requirement. This is part of what's attractive about condemning another's action as wrong, since we want to have this sort of moral authority over those with whom we disagree. And while I admitted up front that not every instance of a duty shares all of these properties precisely (perhaps some instances of a duty are not correlated with a right and so don't license a standing to demand, and there are likely instances of "blameless wrongdoing"), these features generally help to explain what is special about obligation. As one of my friends put the point, there is a spectrum of "shouldiness" (Oughtiness? No, that's really not better . . .), and duty sits at one extreme end. It is what you not only should do but what you must do—what you're morally required to do, and so the rest of us have a stake in your doing it.

Claiming that something is a duty, then, or violating a duty, is serious business. When I make a promise, I bind myself to another in a significant way, since failing to keep my promise opens me up to all of these moral consequences. The person to whom I promised may well demand that I do what I said I would, and the pull I feel from that demand is due to it being legitimate; I know that they have the standing to require that of me. Further, that means that if I continue to fail to keep my promise, I owe an apology and/or compensation, and even with that I may face interpersonal consequences, such as blame and punishment. And those consequences would be justified. The moral seriousness of this situation is why I often decline when my daughter asks me to promise her something that I said I would try to do (get her ice cream, say). If there is too much uncertainty involved in my ability to do the thing she wants, I refuse to promise, because promising is serious business.

These features of duty and obligation—the invasiveness of this component of morality—are why I agree with Maggie that you are not obligated to have sex with anyone. Because sex is not demandable. I'm sure most people reading this book probably already know this, but here's a note especially for younger readers (though perhaps even some older adults could benefit from hearing it): you don't owe sex to anyone. I know, this might make me sound a bit like your dad or your sex-ed teacher. But I think this claim is deeply true for interesting, sophisticated reasons relating to how morality works. Access to your body is something that you get to unilaterally determine, for any reason, no questions. That is why permissible sexual encounters require enthusiastic, ongoing consent. (Good sex requires much more than that, but that's for a different ethics book.)

With this sort of language, we can now provide a nuanced evaluation of other important ethics cases. Consider whether or not you should be a living kidney donor. Right now, there are people dying of end-stage kidney failure, and most of us could save someone's life by donating one of our healthy kidneys. This is not the easiest thing to do, for sure. But it's a safe, routine operation, and if you do it, you will be giving someone else the most incredible gift—years more of life. But not only do most people not donate a kidney; most people probably don't think they are obligated to do so. Are we wrong about this? Should we believe and act as though giving one of our organs is required by morality?

That depends on whether we think our organs are demandable by others. And I do not, in fact, think that they are. In the same way that no one is entitled to your body for sex, no one is entitled to your body for their health. Yes, saving a life is more important than sex, but the authority over your body still stands: no one gets access to your body without your consent. To be clear: Donating a kidney is an incredible act. It is kind, generous, and praiseworthy. But not every act that makes the world better is the sort of thing that is appropriate to demand of people, which is why I don't think you are duty bound to do it.

Maggie thinks that there is a property that holds together some of these actions that aren't typically demandable, and that explains why others are not typically entitled to them. These actions are "intimacies of the first order." To have sex is to share your body with another, giving them access to you and making yourself vulnerable. To donate an organ is to allow another to enter you and to take a literal piece of you. On Maggie's view, such deeply intimate actions enjoy some protection from the demands of morality—they are "matters over which one deserves special moral deference." When it comes to the most personal aspects of our lives, we don't want anyone else to be able to require that we do one thing or another. We, and only we, should get to decide something like who is able to touch our bodies.

By appealing to intimacy rather than something like "bodily integrity," Maggie leaves the door open to other sorts of intimate decisions also being protected from the most demanding aspect of morality. So, for instance, she also thinks that one's decision concerning whom to marry is not typically the proper subject of duty or obligation. Even if marrying a very wealthy person would save your family from bankruptcy, for instance—and you liked the person, and they were good to you, and they generally made you happy—you still wouldn't be obligated to marry them. Forming a family through marriage is a relational intimacy, and just like in the bodily case, it is therefore not demandable by others.

In addition to what may seem like fairly rare, extreme cases (granting people access to our bodies or our emotional selves), I think that the intimacy argument explains much more mundane features of our ethical lives too. If granting others access to our bodies is the sort of intimate act that we tend to have unilateral authority over, then we should expect that fact to be relevant to another big part of our lives, which concerns our interactions with medicine. And that is precisely what we find, as rules governing access to patients' bodies constitute a crucial component of modern medical ethics. Doctors do many things to us that would be clearly wrong if they were done outside of the context of medicine:

they poke, cut, squeeze, burn, break, stab, and touch us in the name of healthcare. If you have ever had a medical procedure done (at least in the United States, but in many other countries as well), you know that you must give your "informed consent" to the procedure, acknowledging that you understand what is about to happen and that you are giving the healthcare team permission to do it. This requirement of consenting to a medical procedure recognizes that access to a patient's body is something that they have full discretion to grant or deny. If a competent patient refuses medical treatment, then it is absolutely prohibited—even if that treatment would save their life. Were a physician to try to force a treatment onto a patient who refused, that wouldn't be medicine; it would be an assault.

We are not then typically obligated to engage in first-order intimacies, as we have special moral deference in such matters. The relevance to the abortion case should be obvious: gestation is a first-order intimacy if anything is. To gestate is to share one's body with another—to intertwine oneself with them. It is to engage in the creation process with every major biological system of one's body and to work toward becoming a biological parent. Part of what Maggie did that was revolutionary in the abortion debate was to point out that nearly all parties to the debate—even Thomson, who injected some much-needed nuance by investigating whether withdrawing life-saving assistance is always a killing—failed to appreciate the complexity of gestation. To think that the moral status of a fetus plays one role, and the rights of the pregnant person play another, is to pretend that we are dealing with a case in traditional morality, with two discrete parties. It is to endorse an atomistic view of a context in which there are no discrete atoms. There is a thorough intertwinement in which one party is effortfully creating and maintaining the life of another developing party in the most intimate way imaginable.

The ethics of gestation, then, asks what we should think about the responsibilities of the creator in such a case. If we tend to enjoy special moral deference when it comes to intimate actions, and gestation is an intimate action, is abortion always permissible?

Maggie doesn't think so, but that's because the question as posed is too vague. We still haven't articulated all of the moral concepts needed to provide a thorough ethics of gestation.

On Maggie's view, some late abortions can be strictly impermissible for a fairly straightforward reason, which was recognized by Thomson as well: the moral right to decline to gestate does not imply moral permission to kill a fetus if it could live outside the womb. Once a fetus is viable, choosing to abort it would entail killing a being that could have been delivered instead. Thus, if a gestating person were to decide they no longer want to be a parent, but they are thirty weeks along, wanting to cease gestation looks very different from when they were twelve weeks into pregnancy. At twelve weeks, ceasing to gestate entails the death of the fetus, and so declining to gestate just is aborting a fetus. At thirty weeks, choosing not to gestate could amount to delivering a baby.

Of course, the vast majority of abortions occur very early in pregnancy, with only about 1 percent of all abortions taking place after week twenty-one. But the point here is important: extensive permission to decline to gestate simply does not entail permission to kill a being that could survive the withdrawal of assistance. Which makes it crucial to answer the question: At what point in pregnancy is a fetus capable of surviving such a withdrawal? It's actually not clear how we should respond, as "viability" (surviv-ability outside the womb) depends on many factors, including neonatal technology and the health and weight of the fetus. So-phisticated neonatal intensive care units (NICUs) are sometimes able to resuscitate babies born at twenty-two to twenty-five weeks of gestation—so early that they won't survive on their own, but have some chance at survival with extensive medical resources. Each passing week of gestation, however, increases the fetus's chance of survival, and it is commonly said that a fetus is viable at around twenty-four weeks in a community with advanced neona-tal technology, though viability in a community with fewer re-sources may be much later. Given all of that, a pregnant person who declines to continue gestating is faced with very different choices at twenty weeks (when the fetus simply cannot survive outside the womb, and so ceasing to gestate automatically entails abortion) and at twenty-eight weeks (when the expectation is that an otherwise healthy fetus can often survive, and so declining to gestate would not automatically entail abortion). In a healthy preg-nancy, even having full permission to decline to gestate does not entail moral permission to abort in the latter case.

Thus, extensive permission to decline to gestate does not auto-matically entail the permissibility of all abortions. But this is fairly unsurprising. The ability to control what happens to your body does not grant you the right to kill another independent being, and viability is the name given to a kind of independence. And so via-bility is an important concept in the ethics of gestation. But this doesn't answer the more difficult question: Does extensive moral deference over intimate actions mean that all abortions before vi-ability are permissible?

This is where Maggie's work on abortion runs smack into her work on what I've called the complexities of moral concepts. Although the moral philosophical literature tends to try to answer questions about the permissibility of abortion by focusing exclusively on whether the fetus has a right to life, Maggie throws a conceptual hand grenade into the topic and blows apart many of the ideas that we tend to run together. She insists that the richness of the abortion debate can only be adequately captured by exploring much more than duty.

The first moral distinction to be made is between having a duty to act in some way and having a reason to act in that way. Although the language of "reason" is perfectly intuitive, philosophers use it in a technical sense, so we should dwell on the concept for a moment. A reason to do something is, formally, a "consideration that counts in favor of" that thing. And considerations are just facts about the world. So we have all sorts of reasons all the time. And only some of them are moral. The fact that there is coffee in the kitchen is a reason for me to stand up and leave my perch on the deck, but that doesn't seem like a moral reason. It's what we might call pragmatic or prudential, since it's just about my own idiosyncratic interests. But the other day, my daughter and I saw a turtle trying to cross a busy road, and we realized that it was almost certainly going to be crushed if left alone to (very) slowly make its way through traffic. I don't know whether there is a moral requirement to rescue any wild animal at risk of harm, but what seems much easier to determine than whether I had a duty to the turtle is the fact that it would suffer and die if it was hit by a car, and that was a reason for me to run out in the road and move it to safety. (For all the turtle lovers out there, I can happily report that the rescue mission was a success, and both turtle and human transport made it out of the road safely.) The well-being of an animal that can suffer seems morally relevant, and so when it counts in favor of an action, we call that a moral reason, rather than merely a pragmatic or prudential reason.

That very quick explanation leaves much out, but it is sufficient, I hope, for making the following point: reasons are a sort of unit of moral measurement. And crucially, they are a "more basic" (think: smaller) unit than duty or obligation—on our "spectrum of shouldiness," they are at the extreme other end. When you are duty bound, you are strictly morally required to do something. But if you have a moral reason to do something, there is just some moral consideration that counts in favor of you doing it. And since there are moral considerations everywhere, reasons come cheap. Right now you have lots of reasons: reasons to continue reading, to stop reading, to donate money to a charitable organization, to go exercise, to get back to whatever work you're ignoring, and so on. And none of those reasons tell us that you are required, permitted, or prohibited from doing anything. They are just part of the world of norms, values, good, bad, and moral stuff around you.

To make it even simpler, here's another way to think about reasons: if you have a reason to do something, all this means is that there is something to recommend the act, or something good about it. If you then do that thing on the basis of the reason, and someone asks you why you did what you did, the reason is your explanation. Why did I run out in the middle of the road as if I had no sense of self-preservation? Well, because if I didn't, a car was going to smash a turtle, and that would cause the turtle suffering (moral reason). Also, my daughter was a witness, and she really cares about turtles, and so saving it made her happy (moral reason). More, acting compassionately toward animals is a good lesson for my watchful daughter (even if running out into traffic isn't . . . hmm, moral reasons in both directions). I explain my action by appeal to the considerations that recommended it.

Returning to the case of abortion, then: the first lesson that Maggie taught me is that duty and obligation are such invasive concepts that certain actions should be granted some degree of protection from their demands. We can think of this first idea as a negative one: certain actions are not typically the proper subjects of duty. But there is a tightly related, positive lesson as well, which is that permissible actions may well still be criticizable. Just because an act doesn't violate a duty doesn't mean it's good. It doesn't even mean that it's not bad or even wrong in some way. It just means that others lack the standing to demand it of you, and you therefore do not owe it to anyone. You can act permissibly and yet be selfish, mean, or a straight-up asshole. If you had good reasons to act otherwise, and you ignored them for bad reasons, then the rest of us may judge you, and I may choose to change how I relate to you. Our ethics are not exhausted by duty and obligation, even though we often act as though they are. To paraphrase Hamlet's comment to Horatio: there are more concepts in our moral world than are dreamt of in most people's philosophies.

Returning to our examples of sex and marriage: I've followed Maggie in holding that neither is the proper subject of a positive duty, which is to say that you are not typically obligated to have sex with or to marry some particular person. But that doesn't mean that declining either is always "morally fine." Turning down sex with someone because of their race is permissible, because you don't owe anyone sex. But it's also racist. The color of someone's skin is a terrible reason not to have sex with them, and it says bad things about your character. It is perfectly possible, then, that you permissibly do something that is morally criticizable.

We can imagine something similar with the marriage case. You don't owe anyone your hand in marriage, and yet if you refuse to marry a particular suitor because they are not as thin as you prefer, this reveals some shallowness of character. You are certainly entitled to say no, but you are also criticizable for refusing for that reason.

It turns out that you can be racist, shallow, and likely a variety of other bad things for acting in ways that are completely within your rights.

The crucial lesson of these cases for moral philosophy is that there is a distinction between what we all have moral reason to do and what we are morally required, obligated, or duty bound to do. Moral reasons come cheap: if an act would make someone happy, that's a moral reason to do it, because others' happiness matters morally. If an action would help someone, that's a moral reason to do it, as helping is a moral good. Having sex with someone, marrying them, or even gestating a fetus, can be acts that are favored by moral reasons. But reasons, by themselves, don't add up to obligation—not even when there are a lot of them. When there is an overwhelming number of reasons to do something really good, we might be comfortable saying that it's the recommended course of action, or even that it's what I ought to do. But even that "ought" doesn't come with the conceptual baggage of duty. If giving someone a book would help them (and I have no use for the book), then I have a good reason to give it to them, and perhaps even ought to give it to them. But if I promised the book to them, then I am obligated. In this case, they are entitled to the book and have the standing to demand that I give it to them. Without the promise, it's a gift. With the promise, it's owed.

In addition, when I act for some reason, that action has a particular meaning as a result. Handing over the promised book begrudgingly means one thing (yes, I did my duty, but maybe I'm also a bit stingy); whereas refusing to hand over a book that wasn't promised can also reflect badly on me—if, for instance, I keep it only so that you don't do as well on an upcoming exam, which will make me feel better about myself. In this latter case, despite not owing you the book, failing to hand it over is mean-spirited and could justify you changing the way you relate to me.

Our first conclusion of all of this reasons-talk, then, is that gestation can be favored or disfavored by reasons, and that how someone acts in the face of these reasons gives us a richer picture of morality than simply reporting whether they had a duty to continue gestating or not. But Maggie thinks there is even more to say, because the value of early human life generates different kinds of reasons.

On a classical picture of moral reasons, the idea that they "count in favor of actions" makes them not only the sort of thing that can justify action, but also the sort of thing that one can be criticized for ignoring. Reasons add up to stronger moral claims. If I have a reason to save a suicidal turtle, and no reason at all not to, then saving the turtle is what I have most reason to do, and so what I ought to do. If it's a moral reason, then it would be wrong not to save the turtle in the face of such a balance of reasons.

But not all theorists think this is how reasons work. Jonathan

Dancy, for instance, thinks that some reasons can "entice" one to act without it thereby being the case that one owes any justification for not acting. Maggie is also in this camp, having argued for the existence of what she calls "commendatory" reasons, which commend an action but do not thereby make it the act that one ought to or must do. Combined with the foregoing exploration of duty and obligation, then, we can articulate at least three categories of moral reasons that can be at play in different contexts.

1. Juridical reasons—named after Kant's term for strict duties—are reasons generated by duty. My duty to keep my promises grounds a juridical reason to take my daughter for ice cream, having promised that I would. Others have the standing to demand that I do what I have juridical reason to do, and so the intimacy argument from earlier implies that I tend not to have such reasons to perform intimate actions. Each of us has special moral deference that prevents the existence of intimate juridical reasons.

2. Ought-making reasons are what philosophers sometimes call peremptory reasons. These are reasons that demand justification for choosing against them, and so can add up to "oughts" or "shoulds." In fact, Maggie thinks that peremptory reasons can make an action morally wrong even if there is no duty not to do it. What this reveals is that we may want to morally criticize an action as violating justified norms but still not take it to be demandable—perhaps for intimacy-based reasons.

3. Commendatory reasons highlight that there is something good about an act without thereby requiring a justification for declining what action is commended. This category of reasons breaks the link between "there is something good about an act" and "therefore you must either do it or have a justification for not doing it." Commendatory reasons preserve moral latitude by allowing agents to decline commended actions for any reason or no reason at all.

On Maggie's view, all three kinds of reasons are relevant to the abortion case, and so a complete accounting of gestation ethics requires a look at all of them.

Early human life has value, she thinks, even when it does not have full moral status. And that value grounds different kinds of reasons for different people experiencing pregnancy. For some, the value of early human life in the first weeks of pregnancy does not count in favor of promoting that life. For the woman who doesn't want to be a mother, for instance, respecting early human life may count in favor of declining to gestate. And while there may be virtue in being open to creating life and becoming a parent, this sort of reason is not in the requiring business. Such reasons are commendatory, and one who finds themselves pregnant can ignore them for any reason or no reason. In this way, not only does Maggie hold that early abortions are fully permissible—she thinks they can be decent and honorable, as they can be precisely the right way to respond to the reasons generated by the value of early human life.

As the fetus develops, the moral situation changes. Maggie is a gradualist about moral status, and so she believes that the kind and weight of reasons that are grounded by the interests of the fetus change over time. As a result, her view preserves the ability to make strong claims about possible moral failure by pregnant individuals who abort late for bad reasons. One way to understand this is that the moral status of the fetus can eventually generate ought-making reasons, which means that they demand justification for ignoring them. If someone tried to get pregnant, then allowed a wanted pregnancy to develop well beyond the early weeks, but changed their mind late (but prior to viability) because they became fearful that they'd never lose the baby weight, this rationale may not justify ignoring the reasons grounded by the interests of the fetus. On the view laid out so far, it's at least possible that one ought not to abort in such a case, and perhaps it would even be wrong to do so if it sufficiently fails to recognize the value of developing human life. But gestation would not therefore be demandable by the moral community, as the fact that it would be wrong to abort does not mean that one violates a duty by aborting.

Which brings us to the category of juridical reasons. If the intimacy argument is correct, then gestation is protected from demandability by special moral deference, which means that a pregnant person has no juridical reason to gestate. Maggie is careful in her writing to use qualifiers when describing this deference, saying that intimate actions "tend" not to be the subject of a duty, and so individuals have near unilateral authority over their own bodies. This seems to imply that there may be extreme cases in which one could be duty bound to engage in an intimacy, though she gives no indication of whether this sort of case might realistically come up in the abortion context.

I began this chapter by criticizing common but simplistic

answers to the question "Is abortion permissible?" And we have explored what I take to be the most sophisticated, nuanced attempt to answer that question, put forward by Maggie Little. But her response is not easily summarized. The idea that pregnant individuals are not typically obligated to gestate combines with an account of moral status and commendatory reasons to generate the view that early abortion is universally permissible and widely decent. However, the proper way to value early human life changes as the fetus develops, which means that later abortions could well be morally criticizable, and even wrong. Once the fetus is viable, abortion could be impermissible, since permission not to gestate does not imply permission to kill a fetus that could survive withdrawal of gestational assistance.

Unfortunately, this collection of judgments does not fit into a pithy slogan.

The case of abortion is obviously important in its own right, but that's not why I have discussed it here. Abortion ethics is not Catastrophe Ethics (as I've defined it). Rather, we've taken this detour because what I learned from Maggie about abortion changed my view about how moral philosophy is to be done. The naïve perspective I carried into my training under her is that ethics is largely about obligation and duty and the myriad other judgments we make about right and wrong, what we ought to do, praise and blame, and (crucially) reasons are either secondary or derivative from that all-important discussion of duty. I typically asked a very simple question about cases of interest: Would it be wrong to do that thing? And I searched for yes-or-no answers.

Our moral worlds are much richer than that, though. And

abortion is the case that opened my eyes to the complexity of rigorous ethical evaluation. I won't, in the rest of our investigation here, require that we perfectly understand and adhere to some particular view of the moral world—neither Maggie's nor anyone else's. I myself don't know exactly what I think about various details concerning the nature of reasons, but I am convinced by the exploration in this chapter of two things. First: that the moral category of duty and obligation is only one part of morality, and so a full accounting of our moral experience will require more than that. And second: that the concept of a reason is a big part of the rest of that story.

CHAPTER 11

(Dis)Solving the Puzzle with New Tools

Although it may be a neoliberal myth that individual decisions have ultimate power, it is a defeatist myth that individual decisions have no power at all.

—Jonathan Safran Foer, *We Are the Weather: Saving the Planet Begins at Breakfast*

"Do you want to stay out in the fields for a bit and watch the fireflies?" The question was posed by my dear friend Liz Brownlee, who, along with her husband, Nate, also happened to be my host for the night at their home, Nightfall Farm. "If we do evening chores after dinner, it'll probably be getting dark by the time we finish," she continued. "We wouldn't have to wait long if you want to see them."

I wasn't entirely sure why she was so excited about fireflies, but it sounded lovely, so I readily agreed. We ate a nice meal consisting of chicken that they raised themselves, alongside vegetables they got from other farmers who sell at the market they attend. After dinner, we tended to the meat chickens, laying hens, and turkeys before forming a makeshift conversation pit out in the field on overturned buckets and bales of hay to wait for the sun to fully set. Tilting her head toward the henhouse where we had just tucked the birds in for the night, Liz looked at me conspiratorially and whispered, "Hey. Listen to this . . ." And then she sang, in a strange sort of high-pitched Muppet voice, "*Me me me me me me me.*" To which the previously quiet (sleeping, I had thought) hens suddenly responded with a chorus of their own "*me me me me me me.*" I burst out laughing, as did Nate and Liz.

I love the farm animals at Nightfall. My favorites are the sheep, whom I had met earlier that day and would get the opportunity to play with again the next morning. They all had so much more personality than I thought possible. The Muppet-hen half-asleep call and response just sealed the deal.

As the sun fully sank behind the farmhouse, I looked toward the darkness in the east and began to understand why Liz loved the fireflies so much. They were *striking*. I had spent my childhood chasing fireflies in Indiana, so didn't expect much from the show. But I realized as I watched my friends' field that I hadn't even noticed how few there were in most places these days. In my own front yard in Maryland, my daughter would occasionally have a handful of them to chase and study; but nothing like the yardfuls I had grown up with. And those childhood memories didn't hold a candle to what I was watching now.

The little lights from the bugs' internal chemical reactions were so plentiful, filling the dark backdrop with brilliant flashes, that it resembled a spectacular light show. I felt like I was at a concert, witnessing the results of some technician's hard, deliberate attempt to dazzle me with the display. The lights danced above the pasture grasses, and we sat for a long time just watching them, enjoying the experience and the time together.

As we sat in silence, watching, Liz almost whispered, "The land is coming back to life. That's what the fireflies tell us." And that's when I realized that the fireflies weren't only beautiful. They had meaning. She continued, "Fireflies are bright little indicators that there is a functioning food web returning to this field. Their larvae eat snails, worms, and slugs, and the adults are often pollinators. The fact that we see so many fireflies means that there is a ton of life we don't see, but it's here thriving as well." She paused. "We brought this field back, and the ecosystem is returning with it. The fireflies mean it worked."

That was the beginning of a long night of talking about farming and ethics, values and commitments. I came to Nate and Liz because they are two of the smartest, hardest-working, downright *best* people I know, and they have chosen a very difficult life for themselves. They've chosen to be regenerative farmers on a small family farm in southern Indiana, where they'll never be rich and they'll always work hard. I wanted to know why.

I met Liz and Nate at Hanover College and have known them since before they knew each other. I watched them grow together as a couple, and I had the genuine honor of officiating their wedding ceremony (on the same family land that would become Nightfall Farm a few years later). And I watched them decide to become farmers—slowly, over years, apprenticing at farms across New England, learning everything they could, and becoming ever more passionate about the land. They didn't just want to farm, though.

They wanted to use farming to heal the land and feed people good food. They wanted animals rather than chemicals to fertilize their fields, and they wanted those animals to have good lives. They wanted to rebalance their little corner of the world, producing food without destroying the planet. They wanted the fireflies to come back.

I remember a discussion with them during a visit right before they started the farm. They had put together a business plan, and Liz's parents had agreed to let them take over part of the family farm (not that her folks were thrilled about the idea—they had stopped farming decades before, when shifting agricultural policies required small farms to "get big or get out"—but they were willing to let the kids have a go at it). We talked about their business plan, and they told me they hoped to break even, financially, by year five of operation.

I was suddenly very concerned for my friends. They were about to work sunup to sundown, seven days a week—giving their bodies and their youth to the land—to operate at a loss for half a decade.

Now when I came to visit these many years later to play with the animals, watch the fireflies, and shoot the shit about farming and ethics, they were thriving. They had just taken over another forty acres of land to continue their expansion, and they sold out of meat shares every year. They had two young apprentices working with them, and Liz spent much of her non-farming time doing regional nonprofit work and environmental education. They put so much good into the world, and so very little bad.

When we got back to the house from the fields and continued our discussion, I asked them bluntly, as only close friends can, why they do what they do. "You work physically hard jobs," I started.

"All day every day. You get into the house late, exhausted, and Nate's over there"—I gestured to my left—"checking his beard for ticks. You did this for years without making any money, and you still won't get rich off of it. You don't have to work this hard. You could do other things. Why do this?"

They smiled at each other and took a beat. "Do you wanna go first? No, yeah, you go." They talked over each other. It was cute. This clearly wasn't the first time they had thought about such things. Nate kicked things off.

"We'll probably give you different answers," he said, "but I can start." He gestured at Liz. "She's changing the world. She's the large-scale change person. I'm the small-scale change person." Nate told me about everyone he really admired in his life—coaches, authors, professors—and how they were doers. When there was work to be done, they did it. Although he didn't use the language, he seemed to be talking about character. Integrity. Living one's values and being part of the solution rather than the problem.

Liz, by contrast, talked about norms, culture, and even politics. People have to change the way they interact with the environment, with the land, and she was desperate to be part of that change. Sometimes when she spoke, I heard the language of being an example—almost a proof of concept that there's another way to live. And other times, she seemed to lean on the ability to influence people—to be part of a movement and to grow its membership.

After a while of just sitting back, I asked the question I had come here to ask. "So do you think you're making a difference? Like, a difference to the world?" They both laughed.

"We're not deluded," Liz responded. "I mean, we can make a difference to our animals, and our land, and our community. But we're not reversing climate change or environmental degradation here." It was interesting to hear them immediately understand the force of the sort of causal impotence that most of us have in the face of these massive collective problems, and simultaneously to not have that undermine their moral motivation at all. "You know, Nate had a great response to the challenge that what we do doesn't matter," Liz said. "Someone recently told us that regenerative farming can't feed the world. He responded that feeding our community is part of feeding the world."

"And what about you, Liz?" She seemed to care more about making an impact, and so I was particularly curious whether it bugged her that the problems she cares so deeply about are too big to get her arms around. "Why not just do something easier, if you can't fix the environment?"

She smiled a little mischievously. "Because living like we do is an act of resistance."

Nightfall Farm is a moral endeavor. Liz and Nate are doing what they feel called to do for moral reasons. But interestingly, none of their explanations for why they feel they should do this entail the conclusion that they, or anyone else, are required to do this. In fact, when I asked them about it, they quickly denied that anyone has a duty or obligation to live like they do. The language they used instead invoked what I think of as "softer" moral concepts, such as integrity, solidarity, participation, and character. Unsurprisingly, philosophers have invoked all of these concepts in an effort to solve the Puzzle.

Recall that when we discussed virtue as a potential explanation of why you shouldn't joyguzzle, for instance, the problem I raised was that it simply wasn't clear why possessing any particular character trait should entail that you have a duty not to joyguzzle. The reason this is relevant is because the motivating intuition behind the Puzzle is that it seems wrong to do things like joyguzzle, and so we reached into our often-too-shallow toolbox of moral concepts and tried to explain that with the language of duty. Could the virtue of, say, moderation or temperance really explain why it's wrong to take a joyride in your Hummer at this particular moment? Well, since your single ride doesn't hurt anyone or contribute meaningfully or unfairly to environmental waste, it's not clear why it would.

One response to this sort of argument is to ask why we ever tried to connect character and obligation in the first place. Virtues and vices are powerful moral tools precisely because they capture a category separate from duty and obligation. Rather than tell us which acts are prohibited, permitted, and required, virtue language tells us which character traits we should adopt. Virtue and vices, as the classic saying goes, don't tell us what to do; they tell us who to be.

The philosopher Dale Jamieson has argued at length that in the era of climate change, we not only need virtue ethics to help us understand our moral place in the world, we need a specific set of virtue tools. He calls these "green virtues," which identify morally good character traits particularly in an era of scarce environmental resources. While this may include some traditional virtues such as temperance (avoid excess in your use of precious resources, live simply), it likely also requires new virtues that may not have been required prior to the Earth being cooked by human activity. One of these that I think about quite a lot is the virtue that Jamieson calls mindfulness, which opposes thoughtless overconsumption and waste. This is a character trait that involves focusing on the costs of one's actions and lifestyles and understanding our role in destructive practices. As Jamieson puts it, "A mindful person would appreciate the consequences of her actions that are remote in time and space. She would see herself as taking on the moral weight of production and disposal when she purchases an article of clothing (for example). She would make herself responsible for the cultivation of the cotton, the impacts of the dyeing process, the energy costs of the transport, and so on." Note that this description does not state that a mindful person never buys new clothes (or even never buys new clothes that aren't strictly necessary); rather, a mindful person understands her role in harmful structures and takes on responsibility for playing a part.

During our discussion, Liz pointed to her new rubber boots, which she had ordered for farm work. They were made and shipped using fossil fuels, and she didn't know who and what she was supporting with her dollars when she ordered them. But they made her work easier. So she ordered them, but she clearly appreciated the moral cost of the product, because she brought it up of her own accord. It wasn't that Liz and Nate never participated in harmful or damaging institutional structures; it's that when they did, they understood that they were participating in harmful systems and took responsibility for it. The boots were worth the splurge, given the amount of work they do; but the rest of their (very) well-worn clothes (they called them "broken in" and "comfortable") looked like they had been manufactured at least a decade ago. Being mindful in this way means taking the costs of certain products and activities seriously as reasons to refrain.

Another moral concept that falls short of duty but that seemed at play in my discussion with the Brownlees is integrity. Integrity is variously defined by different philosophers, but it seems, at its core, to have to do with an alignment between one's values and actions. To have integrity is to walk the walk, so to speak. This is the moral concept that explains why we think people should "live out their commitments." While a simplistic account of integrity may seem to directly require that one not act privately in ways that run counter to their values (to be vegan if concerned with animal welfare, to live low-emission lifestyles if concerned with the environment), it's not that straightforward, since there are different ways to "live one's commitments." It is precisely the force of the Puzzle that such individual actions don't in fact seem to run counter to the relevant values, so perhaps integrity actually requires participating in political action, organizing, writing, and speaking about one's passion project, or contributing to some other collective effort. What would violate integrity, then, is being an environmentalist and yet not participating in *any* broader, collective environmental fights. An interesting aspect of my discussion with Liz and Nate is that the large-scale change/small-scale change distinction between them captured well these differing ways of understanding the pull of integrity. For Nate, someone must heal the land and feed people, so why not him? But for Liz, her lifestyle is partially a political statement, which she intends for broader communities.

Whereas integrity is a positive value, explaining the "pull" of being a coherent, moral self, there are also negative values associated with being a part of morally problematic activities. One is the idea of complicity, or of helping, aiding, enabling, or even just participating in some morally bad project. The classic cases of complicity involve activities such as being a getaway driver for a criminal enterprise, or keeping someone's terrible secret, and thus playing an active role in wrongdoing. The cases that give rise to the Puzzle seem to share the key feature of those cases, which is the idea that if one joyguzzles, or eats animal products, or throws their waste in the landfill instead of the compost bin, they are somehow now a part of the bad that is being done. Since climate change and other massive, structural problems are not shared projects that one intentionally aids, contributing to these problems is not the same as being a getaway driver; while you may "be a part" of climate change when you emit, you are not intentionally participating in a joint effort to bring it about. For this reason, the philosopher Christopher Kutz has suggested that such actions are complicit in a related sense: when you contribute to problems like climate change that no one intends to bring about, you don't strictly participate in a collective effort, but you "quasi-participate" in a non-intentional collective action. Perhaps, then, buying rubber boots—even for morally good reasons, like to work on one's regenerative farm—makes one complicit in the harms of climate change via quasi-participation in the overextraction of fossil fuels.

On my view, there are two key takeaways from the debate concerning the Puzzle. The first one is short and sweet, and that is that our Shakespearean lesson from the last chapter is crucial to understanding our moral response to it. The Puzzle strikes us as puzzling because we start with the assumption that there must be an obligation not to do something bad. And that often is our background assumption, because our moral intuitions tend to be overly simplistic. Duty is simple and forceful and lends us moral authority in debate with others, so we call on it whenever something doesn't feel right. But not everything bad is a violation of duty, and just because we can't thump our interlocutors over the head with the big guns of moral language doesn't mean there is nothing to say. Just as Maggie Little showed in the context of abortion: permissible acts can be bad, vicious, dishonorable, and it may well be the case that we should do otherwise even if no one can demand that we do so. It looks to me like it may be permissible to contribute to massive systematic harms like climate change, but that doesn't mean that it's perfectly OK to do so.

The second lesson, then, is that we need a way to account for the moral pressure we often feel, even when we don't have a duty or obligation to act. And for that, we need to go back to the idea of a moral reason.

As I noted earlier: one of my deepest commitments regarding moral philosophy is that we pay too little attention to moral reasons and focus far too much on duty and obligation. We talk almost exclusively about the most invasive moral concept, which applies in only the easiest cases, but we ignore all the building blocks of morality. Recall the daredevil turtle from the previous chapter: if I had ignored it, leaving it up to the whims of fortune whether it safely made it across the road, I don't think that would have been wrong of me. But there were all these good things about saving it, and they cry out for an explanation of inaction if I choose to walk away. So if I tell my daughter not to worry about it, and we just move on, it's important to ask why. If the answer is that I really thought it was dangerous, then that seems like a pretty satisfactory response. The moral reasons to rescue the turtle asked me to justify a decision not to do so, and a concern for safety is a pretty good justification. But if the answer is that I just didn't want to stop because I'm lazy, now it looks like I failed to rescue the turtle for a pretty crappy reason. It still doesn't mean I violated my duty, but something morally interesting happened. This is the kind of action that I take every day, and that is the foundation of who I am as a person and how I try to respond to the moral world around me. These actions become habits and habits become character, and other people are likely to relate to me (or not) based on my habits and character. This is not to blow out of proportion a single incident with a turtle; rather, it's to show that even if it's a minor encounter, it's not a nothing encounter. How we respond to moral reasons matters. After all, reasons recommend actions.

I think everyone who has tried to explain why you shouldn't joyguzzle has likely failed to establish a duty not to joyguzzle but has given us good explanations of the many reasons why you shouldn't joyguzzle. Doing so is a failure to live out environmental commitments; it exhibits a lack of temperance and mindfulness; it makes you a part of the problem rather than the solution. Really, this last bit is what strikes me as the most intuitive reason not to joyguzzle—it's about what we choose to align ourselves with and become a part of. We may not meaningfully contribute to climate change with our very many emitting actions, but we do thereby participate in it. And when we choose the anti-environmentalist option, we decide not to do our part—however small that part is.

This "contributory" or "participatory" part of ethics is not foreign to us. My mom certainly taught me as a kid to "do my part" and to participate in group efforts. It's just that such advice seems to lose its pull a bit when one's part becomes very small. And as I admitted: the scale and complexity of the problem can, indeed, make it the case that you are not obligated to do your part. But that doesn't mean that there is no longer anything good or bad about what you choose to participate in. Your vote almost certainly will never decide an election; but participating in democracy is important. And even if your work isn't critical to the company, if you choose to work for an organization that is aimed at injustice or evil, being a part of that effort is bad. In the same way, emitting greenhouse gases makes you part of a terrible, massive problem, and that's a reason not to do it. Deciding to buy an efficient vehicle or to invest in renewable energy makes you part of the solution, which is a reason to do it. These reasons often won't be determinative, but they are real. They make up what I have come to think of as the participatory component of ethics.

Thinking about the value of how we participate in various collective efforts as an important part of morality is the key move here. It's good to do our part in pursuit of some collective effort, and that goodness (or value) grounds a reason. Thus participatory reasons are easy to justify; even Sinnott-Armstrong (and his later coauthor Ewan Kingston) explicitly acknowledged that a lack of a duty not to joyguzzle does not mean that there are no reasons not to do so. Rather than responding by trying desperately to rescue a duty from the jaws of causal impotence (as many philosophers have done), I think we should be content with having reasons not to be a small part of large, structural problems. But then we should go a step further and note that these reasons are a very important part of our moral lives. Identifying moral reasons isn't a consolation prize for ethics. This is often where the real work is done. Duty and obligation are rare compared to the prevalence of reasons. So thinking about how we ought to act and live will more often be an exercise in responding to recommendations and what is good and bad about ways of being than it is about conforming to duty.

We are now in a position to solve the Puzzle—or rather, to dissolve it. When we contemplate contributing to massive moral problems like climate change, it seems to many of us that there is something wrong with such actions. You shouldn't joyguzzle, you shouldn't throw food waste or cardboard into the landfill bin when there are compost and recycle bins right next to it, you shouldn't leave lights on when you leave the room, and so on. But if none of these actions actually make a difference to the environment, it seems like you aren't required to refrain. But both intuitions can be accurate at the same time: you aren't required to refrain, but there is, indeed, something wrong with those actions—namely, that there are good moral reasons not to perform them. The Puzzle got its grip on us by exploiting our overreliance on the concept of duty and our tendency to see it as the only moral tool at our disposal.

It's important to note that this solution to the Puzzle isn't an unfortunate compromise. It's not the case that what we should really want in these circumstances is to identify a duty but—seeing as how we can't justify one—we'll have to settle for mere reasons. To see this, let's return to our understanding of duty as second-personally invasive. In the context of abortion, this was important because Maggie Little argued that some intimate actions are not the sort of thing that is demandable by others. But I think the Puzzle shows us that there are other contexts in which duty seems like an inappropriate moral tool, which concerns those actions that contribute to bad outcomes but don't make a meaningful difference. Most serious environmentalists I know still fly for fun occasionally. They take their families from the United States to Europe or Mexico, for instance, and they may semi-regularly drive to the beach. These outings generate "luxury emissions"— greenhouse gases expended that weren't necessary for one's job or basic needs. And yet, I don't think I, or anyone else, have the standing to demand that these folks never treat themselves to a vacation. Yes, we all need to fly and drive less, but their single vacation isn't going to change anything, so it seems unreasonable to expect that they refrain. Duty and obligation are invasive because they are second-personal, and when your action doesn't causally matter, that undermines the justification for the rest of us feeling entitled to demand it of you.

To be clear: this is not an application of the intimacy argument. Taking a vacation is not intimate in the way that sex and gestation are. Rather, in both intimate and causal-impotence cases, I'm suggesting that the invasiveness of morality is inappropriate. It's inappropriate in the case of intimacy even when someone's actions could cause significant harm and benefit, because we are entitled to some degree of moral deference about the most personal aspects of our lives. In the Puzzle cases, the stakes of positing invasive moral judgments are lower, but it still seems inappropriate given the fact that such actions don't measurably contribute to the collective problems. In other words: we lack the standing to demand that others withdraw from participating in all collective harms precisely because those actions don't make a real difference.

That does not mean, however, that we should never judge others in any way for their small contributions to massive collective problems. Part of what seems permissible about the environmentalist taking their family on vacation a couple times a year is that this is a splurge, counter to the backdrop of privately or publicly engaging in environmental work. But when someone gives absolutely no thought to the cost of their actions and builds a lifestyle around gratuitous emissions, we start to see their actions as part of a problematic pattern. They are ignoring real reasons, without good justification, and that is morally bad.

In 2022, the sustainability marketing firm Yard created a report that detailed the flight habits of celebrities, and the results were shocking. According to the report, some celebrities flew hundreds of times over the course of less than eight months, with some flights lasting less than twenty minutes. In response to the report, many environmentalists were outraged, and it's that moral emotion that I think is interesting in this context. No, a seventeen-minute flight will not significantly worsen climate change. But the lifestyle of using private jets for everyday minor transportation is an egregious overlooking of the reasons not to fly. Those who would engage in this behavior aren't mindful, they may seem wasteful and selfish, and we are justified in modifying our relationship with them on that basis. We need not respect, promote the work of, be friends with, or otherwise acquaint ourselves with people who don't properly appreciate moral values. This is how the stakes for responding to problems like climate change can be high, even if we don't have individual obligations.

But note that this is the extreme case. Most of us, as private individuals, try to live responsibly, feel the moral pull of, say, not being wasteful, but also want to have fun occasionally and find it difficult to navigate between those competing drives. Jonathan Safran Foer provides a great example of this in his book *We Are the Weather*, as he argues simultaneously that he and everyone else should eat fewer animal products for the sake of mitigating climate change while admitting that he hasn't yet (even as he finished the book) managed to go fully vegan. If eating animal products were a violation of duty, then everyone in Foer's position should seem to us a bit like a psychopath—knowing full well what's right and wrong, and seemingly not caring enough to act on that basis.

Or, to take another example: in my county, it has become very easy to compost. You just request a green bin online and put your compostable waste in that bin, taking it out for pickup on recycling day. It doesn't get much easier than that. As a result, our family has started composting, and it seems obvious to us that there are good reasons to do so. It's recommended. There's something good about it. And it takes practically no additional work at all. But I don't treat my friends and neighbors who aren't participating as I would if, say, they were habitually lying to me, or carelessly stepping on my toes all the time. They aren't violating rules that I'm entitled to demand that they follow. It would never occur to me that their failing to compost, or Foer's inability to cut the last animal products from his diet, makes it difficult or impossible for me to be friends with them.

This is the complex moral situation we find ourselves in when it comes to being an individual in a massive, global, interconnected society. Many of our daily actions matter morally, but they aren't required or forbidden. We may not be able to stand up and demand that others reduce their carbon footprint, but at a certain point, one's inattention to the relevant reasons makes it clear to the rest of us that they're just not very good people. Eventually, if someone doesn't care enough about any of the problems they are contributing to, we might just start to think they're an asshole.

So far, this is only a diagnosis. I've seen the patient—the Puzzle—and I've identified what I think is going wrong. We get twisted up about modern morality because we expect all moral failures to be failures of duty, but that doesn't fit the structure of so many problems we face. A diagnosis, however, is not a treatment plan. At this point, you'd be justified in asking that central question of ethics: So what the hell do we do about it? We aren't required not to joyguzzle, but there is something bad about joyguzzling; there's a reason not to do it, so it's not recommended. So what should I do? What's the positive advice?

This question only gets harder when we recall that climate change is not the only challenge that gave rise to a version of the Puzzle. I argued in Part I that the Puzzle is everywhere. So how are we supposed to organize our lives?

PART IV

CATASTROPHE ETHICS

CHAPTER 12

Everyday Ethics: Rules for the Twenty-First Century

> Since it is not possible to avoid complicity, we do better
> to start from an assumption that everyone is implicated
> in situations we (at least in some way) repudiate.
> —Alexis Shotwell, *Against Purity: Living Ethically*
> *in Compromised Times*

We all have a lot of reasons. All the time. To do a lot of things. And that's because reasons come cheap. We have a reason to do something every time there is a consideration that counts in favor of doing that thing. Is there something—anything at all—to recommend that you do jumping jacks right now? Is there something good about it? Then you have a reason. (The philosopher Mark Schroeder once suggested that you have a reason to eat your car, since it would provide you with the daily recommended allowance of iron.) And that means that we've always been bombarded by reasons. But many of those reasons don't matter very much, and so we're justified in proceeding with our lives without deeply considering them. You ought never to wonder whether you should eat your car; and unless you're considering your exercise schedule, it's generally not appropriate to wonder whether you should stop what you're doing and do jumping jacks. In today's world, however, in which so very many of our private actions implicate us in massive, structural problems, we have a dizzying number of reasons at all times—and many of them seem to really matter. Unlike car-eating reasons or random-jumping-jack reasons, refusing to participate in evil and endeavoring to be part of goodness and justice are moral reasons and seem to be important.

We can't possibly deliberate about all the reasons that face us constantly. Firstly, there are simply too many of them, and we don't even know about most of them. If it seems strange that you have a reason you don't know about, just think of a simple case: I may not know that my favorite chocolate brand employs child labor, but the fact that buying that brand supports child labor is a bad feature of the act—it still counts against supporting the company, even if I don't know that it does. So the deliberation strategy that tries to consider every reason I have for any action is likely impossible in practice.

But it's also self-defeating to even try. That's because the effort, time, and resources that you would have to spend trying to deliberate about all the relevant reasons you have would undermine your ability to actually respond to those reasons. It would be paralyzing, requiring that you constantly research all of the far-reaching impacts of every action you take and then carefully try to weigh those implications against one another. But the reasons you have to do all that research and weighting may well be weaker than the reasons to do other things, like, you know, actually go out and act in the world.

We thus find ourselves in a strange situation when it comes to evaluating all the reasons we have to contribute or not to various good and bad outcomes: trying to respond to them directly will likely lead us to respond badly. Or, to put the point another way: we are likely to do better at responding to our reasons if we don't try to address them all individually.

According to the Paradox of Hedonism, attempting to seek out pleasure directly leads us to a similar fate. We can imagine a hedonist, embracing their hedonistic ways, and setting out to live the happiest, most pleasurable life imaginable. And so at every moment, they choose the path that leads to pleasure. Predictably, this effort will lead to a rather unhappy life. Why? Because pursuing pleasure directly is self-defeating. The most pleasurable lives tend to be filled with accomplishments, deep, meaningful relationships, and a fulfilling career. But those things are famously hard. Success, love, friendship—they all take work. And so someone pursuing pleasure at every moment will choose against them. If your actual goal is a life filled with pleasure, you do best by forgetting that goal and instead aiming at the lifelong projects that tend to lead to true, enduring happiness—even when this requires sacrificing short-term pleasure or taking on significant hardship.

The Paradox of Hedonism is an intuitive case that shows we must sometimes adopt a strategy of not aiming at some goal in order to better realize that goal. Versions of utilitarianism have recognized the same thing, acknowledging that trying to promote happiness (for everyone, admonishes the utilitarian to the hedonist—not just oneself!) can be both practically impossible and pragmatically self-defeating (this problem, cleverly, is called The Paradox of Utilitarianism). So-called indirect forms of utilitarianism, for this reason, acknowledge that we should sometimes use rules to provide concrete guidance as we attempt to reason our way through life. On this sort of utilitarian view, one is obligated to promote the most happiness for everyone; but since trying to calculate the likelihood of happiness from any act is difficult and we are liable to screw it up, we shouldn't try to figure out what most promotes happiness for every single action. Rather, we should come up with rules that, if we follow them, will tend to promote happiness over the long term. A similar move has been used to justify aiming at the development of certain character traits that will, if adopted, more reliably lead one to promote the general welfare.

From the Paradox of Hedonism and indirect forms of utilitarianism, I borrow the idea that a good strategy for responding to an overwhelming number of reasons may be to look away from the individual reasons and instead to formulate heuristics, or rules of thumb, that guide us more reliably to actually responding to those reasons. Our question, then, is what those rules might look like in a world full of large-scale catastrophes.

One area of life where we're accustomed to using (or seeing people use) self-imposed moral rules is when it comes to food: you likely know people who are vegan, vegetarian, pescatarian, committed to organic, free-range, and more. Although some people may adopt one of these approaches to their diet for purely health reasons (perhaps adopting vegetarianism at the recommendation of their physician), all of the ones I listed can be, and often are, adopted for moral reasons. Some folks try not to eat certain kinds of animals because they believe it causes unnecessary suffering; others want to support certain kinds of farming that they believe are morally better. So we can view approaches to food as a good example of moral heuristics, or simplifying rules, that help individuals guide their repeated, daily behavior.

When considering the reasons relevant to eating animals, a natural moral pressure seems to push us toward dietary extremes. If animals matter morally, such that considerations of their suffering favor not torturing or killing them for our pleasure, then we should not eat them. In highly industrialized societies, much of the meat available is the result of so-called factory farming, in which animals are considered products to be efficiently created, raised, killed, and processed. This focus on efficiency leads to harmful processes such as confining animals to incredibly tight spaces; subjecting them to painful procedures like "debeaking" (to prevent chickens from pecking one another when in such close proximity) without anesthetic; restricting movement so as to "marble" their meat; taking the young away from their mothers; and generally depriving the animals of species-specific opportunities for leading a good animal life. And of course, at the end of all of this, they are killed, often in ways that generate fear and anxiety. As a result, any intake of dead animals at least implicates you in a system that kills other living beings; but unless you are very careful about where your meat comes from, it likely also implicates you in a system that essentially tortures those beings. So the rule that recommends itself in response to animal-welfare considerations is at least as strict as vegetarianism: we shouldn't eat animals. Though perhaps the most plausible rule is actually stricter: if dairy, eggs, and other animal products also tend to come from a system that causes animal suffering, then eating yogurt or cheese also implicates us in a system that causes harm. So moral pressure not to participate in such systems may push us toward a vegan diet, withdrawing our support for all animal products.

We may feel similar moral pressure in forming heuristics for regulating our participation in climate change or ethically problematic consumer practices. Emitting greenhouse gases makes us a participant in the collective action that will harm and kill millions of people; so we should minimize our carbon footprints. Buying products from multinational corporations that engage in bad practices makes us a participant in those practices; so we should boycott such companies by refusing to buy from them.

These sorts of heuristics, which tell us to organize our lives around withdrawing from bad systems as much as possible (and ideally, completely), are the rules of an ethic of purity. Seeing the moral considerations of a massive, interconnected, ethically compromised global society as favoring nonparticipation reveals a prioritization of keeping one's hands clean. And it is the most understandable impulse for a morally serious person to have, as it relates to many of the concepts we've discussed so far, such as complicity and integrity. Being complicit, which is a sort of participation in badness, is to be avoided. And failing to live one's values is a bad thing. So for someone who cares about animal welfare to eat animal products seems to make them a participant in a morally bad practice, which reveals their lack of integrity.

The problem with this reasoning is that it doesn't work. At least, not anymore. It may have made sense in a world in which the impact of our actions was more limited, and we had more control over the very structures that govern our lives. But in our global modern world of 8 billion people, whose uncoordinated collective actions are causing untold suffering, virtually every aspect of our lives makes us a participant in evil. An ethic of purity is impossible to satisfy.

Now, focusing on a single domain may make this argument

seem suspect. The vegan, for instance, may object that it's not impossible to withdraw oneself from systems that oppress animals because veganism has never been easier (at least in some parts of the world), so I'm just making excuses for those who want to eat animal products. But this response misses the point. Yes, you can be vegan, but the purity reasoning that pushes you there also pressures you not to emit greenhouse gases. And it also pressures you not to buy anything from flawed companies or individuals. And so on. It's a standard that, when generalized, is impossible to satisfy, because we are embedded in systems that restrict our options for individual choice. Most people in America must drive to be able to work, because of the way we set up our communities; and most cars require burning fossil fuels because of the way we set up our automotive industry. So most people cannot actually choose, in a realistic sense, not to drive internal-combustion engines. As the philosopher Alexis Shotwell puts it: "An ethical approach aiming for personal purity is inadequate in the face of the complex and entangled situation in which we in fact live."

Purity, then, is an unreasonable and unrealistic moral goal. But I think it's worse than that: it's also a morally problematic goal. Despite the fact that moral considerations seem to push us toward purity in the first place, there are at least two serious moral challenges to purity as an ideal.

The first challenge is likely unsurprising, given some of the discussion in Part I of this book. Remember how David Wallace-Wells said that focusing on individual ethics is a neoliberal diversion, and Mary Heglar said that she doesn't care if you recycle, because she wants you in the climate fight? Well, one way to characterize those objections is to think of them as criticisms of a purity standard: focusing on withdrawing your participation from bad systems and structures keeps you focused on individual action as a response to collective problems. This sort of criticism says we should worry less about keeping our hands clean and worry more about how we're contributing to real, collective efforts to fight back against structural evils.

I mentioned earlier that this criticism has a problem: our individual efforts at collective change—just like our individual efforts to avoid participation in collective harms—are minuscule. So saying that we should march in the streets, donate to effective organizations, and volunteer for political candidates who could enact real change rather than focus on our meaningless individual actions seems to be a bit sneaky, as it's not at all clear that I have more impact on fighting climate change or combating animal agriculture by participating in these collective efforts than I do by limiting my participation in the harmful systems. In both cases, I'm just one person with my tiny, private resources, joining a collective effort. But what this criticism does reveal is that a focus on purity comes with opportunity costs for other sorts of participation in collective action. Spending time and resources on limiting my own carbon footprint subtracts from the time and resources I could utilize for various collective climate projects. There are reasons to participate in good efforts (by, say, attending local environmental meetings or phone banking for political candidates with an environmental agenda) in addition to the reasons not to participate in collective harms. Yes, it's good not to be part of the problem. But it's also good to be part of the solution, even when my part, here again, is infinitesimally small. If I join the climate movement and speak, and persuade, and donate money, and phone bank, I still may not be making a meaningful difference to the grand outcome, but I'm playing a role in something good. And since we all have limited time and resources, focusing too much on my reasons not to emit may hinder my ability to respond to the reasons to be an activist.

This sort of tension between purity reasons and other sorts of reasons is especially striking when it seems like we could contribute in a more meaningful way if we focused on non-purity considerations. So, for instance, ethically sourced products tend to be more expensive than their less-conscientious counterparts, since the low cost of many products is the result of ethically bad practices. If a company pays workers a living wage, provides benefits, and is environmentally conscious, that company's product will cost more than that of a company that drives down prices by squeezing its labor force and externalizing costs onto the natural world. But that means that participating in more ethical systems may leave me with fewer resources to do things that contribute more significantly to concrete good. Is it better to spend an extra $100/month on what I think are ethically defensible products or to donate that $100 to a charity that delivers clean water and food to the poorest communities in the world? What about using that $100 to help a sibling or an elderly parent who has fewer resources than I do?

The basic point here is simple, but crucial: focusing on purity comes with trade-offs. We all have a limited set of resources, which means that using some of them to keep our own hands clean limits what we can do to promote other goods.

The second moral challenge to a purity standard is in some ways deeper. It suggests that even attempting to keep one's hands clean, in a world where doing so is an overwhelming project, actually undermines a key part of what it means to be a person and live a meaningful life. And the argument for this view takes inspiration from the twentieth-century British philosopher Bernard Williams.

Imagine that you are an out-of-work chemist, down on your luck. You've been unable to find employment in your field, and the situation is getting dire. Your spouse is unable to work for health reasons, and so you really need to find a job to support your family. Then one day, a friend and colleague calls to let you know that she has found a position for you in a company that designs and manufactures chemical weapons; this friend is in a sufficiently powerful position that, with a good word from her, the job would be yours. However, you are uncomfortable taking the job because you believe that the development of chemical weapons is deeply wrong. Your friend, it turns out, also thinks chemical weapons are bad. She confides in you that the other finalist for the position is just a little too excited about chemical weapons, and this is one of the reasons she is hoping you'll take the job. Working in this sort of position requires a bit of moral seriousness, she believes, and so your reticence is actually one of your qualifications. So what do you think? Should you take the job?

A similar case was originally used by Bernard Williams as part of an argument against utilitarianism. On his view, utilitarianism has a ready answer to this problem and any other like it: of course you should take the job, because it produces good for you, good for your family, it won't produce bad for the world (since someone will be in that job anyway) and may even produce a bit of good (since you may temper the excitement for chemical warfare by virtue of your position). But the problem, according to Williams, is that your moral opposition to chemical warfare just doesn't seem to matter to utilitarianism (at least, not much—if it makes you unhappy to be in your job, then it's one small negative point in the great calculus of happiness production, but it doesn't matter any more than that). As people—as moral agents—we care which actions come from us. We care about the collective efforts in which we participate. It matters to us whether or not the production of chemical weapons was aided by our own efforts.

On Williams's view, utilitarianism requires that we not see what we do as mattering, morally. The world gets organized in various ways through the individual and collective efforts of billions of other people, and then we are required to do whatever generates the most happiness, regardless of our own values. But this way of seeing the world comes with a severe cost: it turns us into tiny cogs in a vast happiness-producing machine. There is no room for individual projects or value structures, since they all must give way to the demand for happiness. The problem with the chemist's case, Williams thinks, is not necessarily that the utilitarian thinks you should take the job; that's a reasonable view regarding a difficult case. It's that they think it's *obvious* you should take the job. That's how little your moral commitment against chemical warfare means to the utilitarian. In this way, Williams famously says, utilitarianism robs us of our integrity. For if you are required to abandon your principles as soon as following them is suboptimal in terms of happiness production, then you don't have principles. *That's* the cost of utilitarianism.

It's a brilliant objection. But it's even more powerful than Williams claimed, as it's relevant to much more than just an evaluation of one moral theory. What Williams's integrity critique really shows is that there is a limit on how far morality should invade into our personal lives if we are to leave ourselves space for the development of meaning. Morality must allow us room for adopting a range of projects, and we need to be able to act according to the values that undergird those projects for them to be meaningful. Of course, not every project is defensible. There is no protection from the demands of morality for being the best torturer or the best assassin. But if your opposition to taking the chemical-weapons job is based on a deep, lifelong commitment to pacifism, then that is the sort of moral project that we want you to be able to act on.

So far, it may seem like the integrity critique is friendly to the purity ethic. After all, integrity was one of the justifications for refusing to participate in a bad collective structure. In the same way that an environmentalist seems required by integrity to minimize their carbon footprint or the animal-welfare activist seems required by integrity to be a vegan, if you are morally opposed to chemical weapons, then you are required to refuse the job in the chemical-weapons company. So a call to respect integrity seems to count in favor of a purity ethic.

The problem is that a purity ethic, when generalized, becomes very much like utilitarianism: it becomes totalizing, or what we can think of as imperialist, invading into every aspect of our lives. What makes utilitarianism integrity-robbing is that it leaves us no room to develop life projects for ourselves. But a purity ethic, in our complex, entangled world, ends up doing something similar. In the case of the chemist, a purity ethic would say that you should not take the job, since doing so would make you a participant in chemical-weapons development. Furthermore, it's obvious that you shouldn't take the job. The rule that tells you to keep your hands clean is simple and straightforward to apply. But that's a problem in cases like these, because they aren't simple. The reasons you have to take care of your family are real reasons, and if you can't get another job and that is putting you at real risk, then you are the victim of structures that have limited your ability to choose freely.

I'm not sure whether you should take the job or not, but I am sure about one thing: the answer is not obvious, and so any account that claims it is has a serious problem.

Like many people who are concerned about animals and the environmental impact of food, I have a dietary rule of thumb. Mine is: eat low on the food chain, especially when it's easy. What this often means is that I eat plant-based foods and sometimes seafood; but if I'm at a friend's house, or with family in Cyprus (which, culturally, has a meat-heavy diet), I don't morally stress too much if my best or only option includes lamb, beef, or poultry. This sort of diet doesn't maintain any sense of purity for me, as my body gets animal protein in different contexts, and so I am contributing to the harms of animal agriculture and climate change to a greater extent than necessary. But I also don't take there to be zero moral cost in eating animals, which is why my rule directs me to reduce it.

This sort of heuristic is similar to one famously offered by Michael Pollan, who advised us to "eat food, not too much, mostly plants." Here, too, we get a recommendation to eat relatively few animal products, and not too much food in general, though the focus on "food" may be at odds with my simpler rule of eating low on the food chain. That's because Pollan differentiates between "food" and "food products," where the former is recognizable as food and the latter is complex and highly processed. What this means is that some vegetarian-friendly foods—like, say, plant-based burgers and sausages—are recommended by my rule for preferring plants and recommended against by Pollan's rule because they are not what he thinks of as real food. But since neither of our heuristics offers an absolute recommendation, eating an occasional Impossible Burger or Beyond Bratwurst will not rule you in or out of compliance, nor will eating an occasional beef burger or pork sausage. In both our cases, there is flexibility and variation.

A similar rule when it comes to my family's carbon footprint is that we should invest in carbon-reducing interventions and technology because we have the means to do so, and the widespread adoption of such technology requires some of the population to opt in. This has meant that we have engaged in commonsense interventions, such as installing energy-efficient appliances when old ones need to be replaced, as well as purchasing electric cars rather than internal-combustion ones when it came time to purchase new vehicles. These sorts of actions certainly don't maintain our family's climate purity, and indeed, by virtue of living in the United States, we still have a relatively high carbon footprint. But the rule recognizes the value of making moves in the direction needed, while not mistaking that moral pressure for a strict obligation (since none of these small actions will make a meaningful difference to climate change by themselves).

In general, what all of these rules do is recognize the real reasons that count in favor of reducing one's participation in collective harms, while simultaneously recognizing that the fact that they're only reasons (and not strict obligations) means that they are more easily responded to by countervailing considerations. Some such countervailing considerations are morally significant:

my dietary norms change in light of cultural norms, which is to say that I take the fact that it would offend my host to be a real moral reason not to reject food offered to me. So meat offered to me in someone's home, prepared by them, is recommended to me by etiquette, which I take to generate a real reason, easily outweighing the moral pressure generated by my heuristic. Other meals relatively high on the food chain also seem genuinely valuable, such as those that come with tradition. The Thanksgiving turkey here in the United States plays a valuable role in many families' lives, and that, too, is a reason to partake. But I also don't think countervailing reasons must be morally high-minded. I sometimes eat turkey sandwiches not for any cultural or etiquette-based reasons, but because they are healthy and good and the alternatives are foods that are unhealthy or that I dislike. A preference for something healthy and tasty is not a particularly good reason to eat something, compared to animal welfare or the environment; but since my individual actions don't meaningfully affect animal welfare or the environment, those considerations can be occasionally outweighed, so long as I generally limit my participation in those harms.

We are not trying to determine what we are required to do. We are trying to determine what is recommended, or what would be good, in a world where our actions seem both to matter and not to make a difference.

Even having abandoned a purity standard, you might reasonably wonder why my own heuristics concern my private consuming and emitting behavior at all. Eating high on the food chain and emitting excess luxury emissions would make me a participant in collective actions that cause massive harms, but focusing on these participatory violations might still feel a bit like Wallace-Wells's neoliberal diversion. After all, if climate change threatens devastation, then some of the reasons we have do not concern our minor contributions to greenhouse gases; they concern our minor contributions to various efforts to limit our collective emissions, to develop technological solutions, invest in climate-resilient infrastructure, and so on. Some of our reasons, in other words, are to join the political and social movements that will actually address climate change.

This is exactly right, though it doesn't vindicate any sort of attitude that the political is *more* important than the personal. One could live a pretty extravagant lifestyle and engage politically to reduce emissions, and there are no grounds to criticize that person and simultaneously give a pass to someone who has a minimal carbon footprint but is not politically engaged. There is no basis on which to prefer one lifestyle over the other, as they are two different ways of responding to the reasons we all have. Jonathan Safran Foer wants you to cut out meat before dinner, and Mary Heglar wants you in the climate fight, but the idea that one of them is the right way to address climate change seems to me to be indefensible. They are both responsive to real reasons and values, and neither is likely to make a meaningful difference on its own. They are both exhortations to participate in goodness rather than badness, even when your own participation is likely to be very small.

The truth, then, is that doing one over the other is more akin to a lifestyle choice than a moral one. The moral choice was to do something rather than nothing about climate change. What, in particular, you choose to do is about your strengths, preferences, and values. The Williams-style argument from earlier suggests that we must have the moral freedom to structure our life projects within limits. Focusing on the political or the private is one such choice that we can make.

Of course, the most hardcore of my opponents is likely to say that this is a cop-out—you have to do it all, the private and the political. Why would engaging in activism buy you permission to emit flagrantly? And why would minimizing your own impact justify lazily letting the world burn without making a real effort to push collective change? But that sort of question only makes sense if you are required to act in a certain way, which we are not. The fact that we're responding to reasons and not obligations undergirds freedom in how to respond.

When you have a reason to do something, you have a justification to do that thing. After all, there is something good about it, something to recommend it. If you choose not to do it, then the only pressure you face is to explain why you didn't do it, despite that justification. Reasons, then, ask us to justify our decision to ignore them. In the case of collective threats like climate change, we have more reasons than we can possibly respond to, and trying to respond to them all threatens our integrity by turning us into machines. This situation justifies our choosing to focus on the reasons that align with our central values, goals, talents, and strengths—by choosing, that is, how to incorporate the problem of climate change into our lives.

The only option that is clearly unjustifiable from the outset is the option of doing nothing. Climate change is a catastrophic threat, and there are millions of ways we can change our behavior in response; failing to change our behavior at all is what we may not do. It's still not a violation of duty to do nothing. You are not morally required, and neither I nor anyone else has the standing to demand that you do something. But failing to do anything at all means that in the face of overwhelming reasons to do a whole multitude of things favored by morality, you simply shrugged and said, "Meh." The moral community would be justified in judging you and changing how they interact with you. They might decide that you're lazy or selfish or have some other vice.

Why do I say "might"? Well, because we have the moral freedom to choose how to respond to the overwhelming number of ways that we can participate in goodness and badness. That freedom applies not just in how we choose to respond to climate change but in choosing which projects to engage with. So if the person who does not modify their behavior in any way in response to climate change has organized their entire life around alleviating poverty in the world, that might change our attitude. Because the presence of poverty is another catastrophe that provides us with overwhelming, but mostly participatory, reasons. Not only can no one respond to every reason assaulting them; no one can respond with the same energy to every problem that the world faces. And so dedicating a certain amount of one's energies to participating in the fight against one catastrophe can be a justifiable way to respond to the fact of catastrophe in general.

This idea that one should do something, but that one has latitude in determining what they should do, is reminiscent of Immanuel Kant's notion of an "imperfect duty." According to Kant, we have perfect duties and imperfect duties, and only the former strictly require a particular action. Having a duty not to lie means you must never lie, and lying once means you've violated your duty. Thus, telling the truth is a perfect duty. But imperfect duties are not satisfied or violated by individual actions, and instead admit of latitude. The paradigmatic imperfect duty for Kant is beneficence, or promoting the good of others. While it's a genuine duty, it does not mean you must, at every instance, promote others' good. You have discretion in deciding how and when you will do so. If you pass by someone who needs assistance without helping them, you haven't thereby violated your duty of beneficence. But you may not pass by every person who needs help. You simply have the freedom to decide when you will discharge your responsibilities to help. Since problems like climate change don't generate duties at all for most of us, my view here is not precisely Kantian. But I think it's helpful to think about the "imperfect" structure. The catastrophes that implicate us all generate strong moral reasons to do something, but we have the moral freedom to decide what, precisely, we should do in response. Recalling Maggie Little's discussion of different kinds of reasons, we can note that the ethics of how to participate in structural change concerns latitude-preserving reasons.

While this freedom—or latitude—gives us permission to choose how we respond, it is important to recognize that anyone who says that we actually should do it all—or perhaps more accurately that it's better to do it all—isn't completely wrong. And they serve an important purpose by keeping us honest. Climate change is among the direst moral threats humanity has ever faced, and so trading off one way of discharging your responsibility against the other always makes you subject to the retort "Why not both?" You could virtually always do more both by reducing your participation in harms and by increasing your contribution to collective action. And while, yes, taking this to the extreme bumps up against the Williams-style moral-freedom critique, couldn't we each do a bit more before we get there? This question reveals the scalar nature of our participatory ethics: we can always be better and worse. We can participate in more justice and in less harm.

And so the friend who pushes us to do both things rather than one, or to add a project to our current list, may well be a better person than we are. That's part of our situation, and it's unavoidable.

The downside of having no duty in the realm of participation is that you can't simply do what's required and be done. There's always more that you could do, and it would be better if you did.

Let's return to the case of Nightfall Farm. I claimed that Liz and Nate put less bad into the world than virtually anyone I know. They also leverage their lifestyle as proof of concept for others, and they educate the public on the land, the environment, and how we can live more sustainably. While they are themselves under no illusion that their hands are morally clean, it's hard to imagine that most of us could do more than them in an effort to withdraw ourselves from harmful practices.

Now compare their lifestyle to that of a billionaire philanthropist. There are plenty to choose from, so take your pick: Melinda Gates, co-founder of the Bill & Melinda Gates Foundation; Michael Bloomberg, founder of Bloomberg Philanthropies; Warren Buffett, who has pledged to give more than 99 percent of his wealth away during his life or at his death; MacKenzie Scott, who since 2019 has given away more than $14 billion and vowed to keep giving "until the safe is empty"; and so on. What these individuals have in common is that they represent something like the polar opposite of Liz and Nate in terms of how they respond to the moral reasons presented to them by catastrophe. Without knowing the intimate details of their lives, I think it's safe to say that they contribute, as private citizens, to harms like climate change to a far greater degree than the regenerative farmers from Indiana. If they enjoy their wealth even a little bit, then their carbon footprints are likely among the largest humanity has ever seen. But the power that their money translates into by virtue of their philanthropic efforts means that they can do unprecedented amounts of good. In the realm of public health, it's nearly impossible to estimate the positive impact that just the individuals named have had over their lifetimes.

So whose life is to be morally preferred? The simple farmers? Or the ultrawealthy philanthropists?

And what about the rest of us? Most of us are, undoubtedly, somewhere between these extremes. I know that I will never have the ability to do the amount of good that Melinda Gates has done, nor am I as committed to limiting my contributions to harm as Liz and Nate are. So what do most of us, rather normal, folks do? Well, we participate in charitable giving and collective efforts more fitting to our means, and we withdraw our participation from at least some harms to at least some degree, but often, many of us feel a bit at sea concerning how we are actually supposed to organize our lives. And yet—and this is the crucial bit—most of us likely know some of these normal folks (and hopefully some of us are these normal folks) that others think of as "good people."

And here is one of the central lessons of this chapter: they're right to think so. A bunch of normal folks are good people. And we're right to laud Nate and Liz. And we're right to laud philanthropists. All of these judgments are right, because there are multiple good lives—multiple good ways of living. The moral freedom we have to decide how to participate in good and withdraw participation from bad means that there is no singular right way to live in our world.

It is not our job to figure out the uniquely justifiable life that we are required to live given the variety of catastrophes we face. It is our job to identify one of the many ways of living a good life—one that aligns with our values, preferences, and even talents and strengths—and then to live it in good faith and with integrity.

You might be thinking that I've given you spectacularly little concrete guidance so far. And you'd be right. But that's because I don't think I can give you much concrete guidance. Responding to participatory reasons requires the injection of subjective values, and so only you can determine what a good life for you looks like. But what I can do is provide some framing for how to go about that process.

Liz and Nate's lifestyle provides them with fairly clear guidance. By thoroughly intertwining their lives with their moral mission, they have relatively few questions to answer about how they should live. Their commitment to the land and environment takes up nearly all of their time, money, and energy. One kind of good life, then, is to become a true champion of an important cause. To make it what you are about. Not only is this how I see Nightfall Farm, but it's how many morally committed people I know organize their lives. They may have many broad moral commitments, but there is one that becomes an organizing principle. Many animal-rights activists I've known over the years have this sort of orientation. They view what is happening to animals as a constant genocide, and so they dedicate their lives to fighting animal agriculture. Veganism is not enough as a response to such an atrocity, they think; so they also orient their volunteer work, their advocacy, and their charitable giving to animal-welfare issues. These are good lives, but they are not uniquely good lives. That commitment comes with trade-offs and opportunity costs, so they are by necessity ignoring lots of other worthy causes.

But since most of us will not be a champion for a cause, this sort of focused organizing principle will not be helpful. That means that most of us trying to wade through our complex moral realities in a decent way are likely to think that lots of causes matter, and that none are unique. And while that's true, there are some causes that are particularly acute. They are catastrophic in scope and truly urgent, meaning that there is special reason to pay attention now. Obviously, climate change is such a threat, and it has yet another property that makes it especially worthy of attention, which is that it is a threat multiplier: it worsens other catastrophes that are independently worth our attention. Racism and xenophobia are worsened by climate disruptions that disproportionately affect the most marginalized and are causing and will continue to cause massive immigration crises. War (including nuclear war) is made more likely by resource scarcity. Animals and ecosystems are threatened by temperature change, wildfire, drought, and more. Infectious disease outbreaks are likely to increase as a result of changing migratory patterns of animals that serve as viral vectors. And so on.

Thus, although there are lots of good lives, and none are uniquely good, there's a strong case to be made that climate change should be among the top considerations for those individuals trying to construct a good life. It is a catastrophic, acute threat. And our lives are utterly intertwined with it, since we contribute to it every day in multiple ways, and there is no longer any excuse for not knowing about it. It's in the news. The weather is changing dramatically. Policies are being debated. Climate change is one threat that most of us—with the possible exception of true champions of other, unrelated causes—should be responding to. It's the paradigmatic catastrophe, so it's worth thinking a bit about how to organize our personal responses.

The framework that follows is a way of generating rules of thumb in response to participatory reasons. I'll use the case of climate change as an example, but the idea is that you could do the same with any other collective harm in which you find yourself implicated. You can start to identify how you want to live your life by answering two broad questions. First, what kinds of reasons do you have in response to the threat? And second, how strong are the countervailing considerations, or the competing reasons?

In response to the first, I suggest that there are three general kinds of reasons generated by threats like climate change. The most obvious are private reasons—reasons to withdraw participation from the harms. In line with the earlier discussion, I will call them purity reasons, though that shouldn't carry a negative connotation—it's good and natural not to want to be part of something morally bad. A purity reason is a "negative" reason, or a reason not to do something. In addition to these, there are at least two kinds of "positive" reasons also generated. The first is what I will call a social reason, or a reason to relate to others in a particular way. And the second is a structural reason, or a reason to engage in efforts to change some institutional, structural, or political feature of the world. These are reasons to go out and do things, and so characterize different ways of being an activist with reference to some catastrophic threat. All of these classes of reasons seem important, albeit in different ways. Purity reasons matter because what we do matters—we don't get the same moral credit for convincing someone else to reduce their carbon footprint as we do for reducing our own. The convincing still matters, though, because our social interactions are how we generate and strengthen movements, and so social reasons respond to the value of being part of something bigger than yourself. And structural reasons matter because actually solving collective action problems requires making changes to structures, systems, and institutions.

In response to purity reasons regarding climate change, one might adopt the rule "Minimize your carbon footprint." This is because climate change will cause harm, and emitting makes you a part of that harmful system; so the badness of participation is a reason to withdraw from it as much as possible, which is to say, to minimize one's contribution. On the other hand, the need for collective action in order to mitigate the harms of climate change counts in favor of adopting various social and structural rules, such as "Talk with friends and family about climate change" and "Volunteer for climate organizations." The value of collective action is a reason to be a part of it, or to try to bring it about.

One thing to note, however, is that these example heuristics are not similar in terms of what they demand from us. Truly minimizing one's carbon footprint is intensely demanding, whereas for at least some of us, discussing the topic with loved ones is not demanding at all (though I recognize that, for some people, discussing climate change with friends and family is a significant burden and can even lead to lasting harm to the relationship). This leads to the second question I said we must answer, which concerns how strong the countervailing reasons for adopting some rule are. At each of the purity, social, and structural levels, we can formulate heuristics that are basically uncontested, because we have virtually no good reason not to follow them. One that I remind myself (and my daughter!) of constantly is the purity-based heuristic "Don't be wasteful." This can be specified in a million different ways, but the most common ones in my personal life concern leaving lights on in the house, cooling or heating to an uncomfortable level, buying groceries thoughtlessly so as to end up throwing them away, and similar habits. While it is true that my reasons not to do these things are modest, given the infinitesimally small contributions I'm making to environmental harms, it's also true that there is basically no reason not to reduce waste. Thus, while it's true that none of the heuristics directing our participation in collective harms or benefits will be genuinely obligatory, some of them seem more optional than others. Even if you are a champion of some other cause, such that you have no real time or energy for the climate, that's still not a reason to leave lights burning. This is part of why I bristle a bit at the idea that people need not change their private behaviors, since "the political is what matters." Yes, there is good reason to phone bank for environmentalist candidates, to volunteer with climate organizations, to donate to green causes, to join climate protests, or even to run for local office. Participating in the collective good in those ways is recommended! But it's not an either/or situation. Going to a march shouldn't replace eliminating wasteful actions or practices. If we want to live good lives, we should care about how we respond to moral reasons, and there's just no good justification for refusing to change some of our behaviors.

There are other heuristics, however, that demand more of us. I think of these as being divided into ones that face modest, self-interested reasons not to heed them, and then those that face genuinely good, strong, often moral, conflicting reasons. The former are where we start to feel the moral exhaustion of so many things that we could do in response to catastrophes, and we just don't want to. Not everyone is comfortable with becoming involved in, say, local politics or joining a protest march. And some ways of trying to be part of a collective change are expensive or take extra energy (think: composting, investing in clean technology). The other category of heuristics, then, is genuinely morally invasive. On the purity front, perhaps we consider minimizing vacations but recognize that travel and exposure to other people, places, and cultures are genuine goods. Or perhaps travel is how one sees their family (say, in Cyprus!). On the social and institutional front, adopting more demanding heuristics involves becoming a true advocate, which—as we've seen—comes with real opportunity costs.

Since each kind of heuristic faces each level of opposition from competing reasons, we can imagine our two questions offering two different axes along which candidate heuristics can be placed, as in the below table:

	No competing reasons	Modest competing reasons	Compelling competing reasons
Purity	Don't be wasteful	Eat low on the food chain	Minimize your carbon footprint
Social	Talk with loved ones	Be vocal within community	Be a loud advocate for change
Structural	Vote for/support climate policy	Donate to climate causes	Volunteer for the movement

In this example table, I've offered possible heuristics, but of course, there are many other options. What I intend to show with it is that most of us, I think, should feel a real responsibility to generate heuristics in the left column, but especially in the top-left, since we engage in so many climate-relevant actions every day. If there is no good reason not to regulate our behavior, then we have very good reason to develop a thoughtful response.

The middle column is a great opportunity for expanding our

moral seriousness—for asking ourselves the hard questions about whether we really should weight our self-interest as heavily as we do, and slowly adding commitments to our moral lives. However, with competing reasons come some that inevitably vary in strength. Those with very few resources (neither time nor money) will find it harder to add political engagement to their lives, and introverts will find various social heuristics harder to adopt. In addition, some people are just not skilled at certain modes of engagement. Perhaps I'd be happy to talk to lots of people about climate change, but I'm just not good at it—I'm not persuasive, or I turn others off, or maybe I'm just quite boring. Part of the freedom to arrange one's moral life involves recognizing that our strengths, weaknesses, and good and bad fortune affect how easy or difficult it is to live according to various rules.

Finally, the right column gets filled in by those who are making the cause in question an organizing principle for their lives. Filling out the entire square (with multiple, variously specified heuristics) is how one becomes a champion of the cause. I imagine that Nate and Liz are champions of the environmental cause. The fact that I can't readily identify any other champions in my life makes it clear how rare it is to truly organize one's life around such a commitment.

Noting the different strengths of competing reasons also makes clear one of the major goals of policy and systemic change, as those structural interventions change how the matrix looks for individuals. Eating low on the food chain, for instance, is more difficult when good, healthy, vegetarian food is more difficult to find. But changes to food availability can make it easier, pushing it left into the "no competing reasons" column. The same is true of minimizing one's carbon footprint, which is demanding precisely because of structural policy decisions that determine the way many of us live our lives. But a decarbonized economy makes it easier to reduce one's footprint, pushing that heuristic left into the middle column. In fact, this is how much of climate policy works—not by mandating certain actions but by incentivizing changes to population behaviors. When buying energy-efficient appliances becomes easy, more people do it. The tax credits that make solar panels more affordable are policy solutions that require some people to now see solar panels as something with weaker opposing reasons. In this way, my individual work advocating for policy change (engaging in actions recommended by the second and third rows) has the effect of changing what I can reasonably be expected to do in the top row. Part of my structural effort is to be part of changes to my purity efforts.

For illustrative purposes, here is how I fill out the table above for myself, concerning my example rules for climate change. In the example below, I use a grayscale to indicate how well I live up to the standards set by each rule—the darker the shade, the better I tend to do at following the rule's guidance.

	No competing reasons	Modest competing reasons	Compelling competing reasons
Purity	Don't be wasteful	Eat low on the food chain	Minimize your carbon footprint
Social	Talk with loved ones	Be vocal within community	Be a loud advocate for change
Structural	Vote for/support climate policy	Donate to climate causes	Volunteer for the movement

As you can see, I'm quite good at following the rules in the leftmost column, but things get a bit scattershot after that. The darkness of the middle row reflects the point about one's strengths and talents playing a key role in filling out such a table. I've invested heavily in responding to the social reasons generated by climate change because I happen to be a scholar and a writer. I can contribute by reaching various audiences in a way that fits naturally with my lifestyle. As a father and husband with a demanding job, however, I find it much harder to give time to the cause through volunteering, for instance. I also find minimizing my carbon footprint among the hardest potential responses—partially because of my interests and values (in, say, travel), but also because I live in the United States, where we, as a society, have made very few decisions that make it easy to live an environmentally frugal lifestyle. Rather than focus on this shortcoming, however, I focus on the areas where I feel better equipped to respond.

This sort of relativity extends to the different stages of life that we all happen to be at as well. Consider the work and the motivating power of climate activist Greta Thunberg, who began protesting Sweden's lack of significant climate action as a teenager. There are lots of elements to her success in sparking an activist revolution, but her age seems crucial: the youth need the planet for longer than those of us who have been around for a few more years, and it's our job to protect them. Young people, then, have the ability to influence the older generation by emphasizing their vulnerability and the fact that the older generations have failed in their responsibility to protect them. Young people also tend to have more time and freedom to engage in activist work than their parents, who may be juggling careers and other obligations, though they have less money and so cannot contribute financially. The youth, then, have good reason to build out their matrix with a focus on activism, using their situation and their energy rather than finances.

Contrast this with older folks—say, retirees—who have time on their hands as well but have a very different set of skills and resources. Climate activist Bill McKibben has recently focused on rallying this group by founding the organization Third Act, which calls on "experienced Americans" in the third act of life to join the climate fight in ways particularly well suited to them. The organization says, "We are building a community of Americans over the age of sixty determined to change the world for the better. Together, we use our life experience, skills, and resources to build a better tomorrow." This takes many forms, but a prominent one is mobilizing the combined wealth of this generation to try to motivate the financial industry to stop funding fossil-fuel projects. In this case, Third Act asks people simply to write to their banks and ask them to stop funding environmentally damaging projects, and to pledge to take their business (and money) elsewhere if the banks do not comply. Generating a rule for oneself, "I will not bank at an institution that fails to live up to certain climate goals," is very different from marching in the street or protesting the opening of a new coal mine, but it is well suited to those with a particular set of resources.

Writers have good reason to write about climate change; farmers have good reason to engage in regenerative farming; kids have good reason to follow Greta Thunberg and others like her—protesting, striking, and gaining visibility; and wealthy people have good reason to use their money to try to motivate change. Joining the climate fight may well look different for everyone, and that's not only OK—it's recommended.

Although this matrix does not tell you precisely how to live a good life when surrounded by catastrophe, I hope it makes clear that what we are really on the hook for, morally speaking, is being thoughtful in how we approach the ethics of participating in collective action. It is intimidating to think that the only good life is one of purity, but it might seem that if we reject the pursuit of purity, we must accept moral nihilism about our efforts—that if you aren't required to keep your hands clean, anything goes.

If there's a slogan for this chapter, it's that we should reject both an ethic of purity and an ethic of nihilism, and instead embrace an ethic of conscientiousness, which directs us to respond to the overwhelming number of reasons assaulting us by organizing them into variously demanding rules and then always striving to live a better, more justifiable life. Since there is no strict duty in this realm, we have latitude in determining how to structure our efforts to participate in the global systems that benefit and harm. What's really important, then, is to be thoughtful about how we do so, recognizing that our participation can be private or public, that it can always be scaled up (but that we can never fully discharge the moral burden these structures place on us), and that being intentional in this way is part of being a thoughtful member of a community. There is no single right answer, but lots of good lives, and so we should exhort others to be conscientious rather than beat them over the head with specific demands or claims about what is required in order to be a decent person.

Conscientiously attempting to set rules for ourselves can help us to respond to participatory reasons as they constantly assault us. These heuristics—which we identify and choose by exploring different, valuable ways of organizing our lives—help us to make justifiable decisions without being paralyzed by indecision over every little thing we do. In this way, heuristics are for the mundane. Food is morally serious, in part because we eat a lot. Weighing out the moral reasons to eat in various ways every time we sit down to consume calories is impossibly inefficient. Our decisions about food morally matter, then, but this is mundane, everyday ethics. The deliberative work should be done ahead of time by thinking about what kind of life we want to live. Then we just live it. The value of a rule of thumb is that we can set it and forget it, simplifying our daily cognitive load.

Some decisions, however, are not everyday or mundane. Some are monumental, shaping the rest of our lives to a significant degree. And with that sort of impact, they are likely to be the subject of many moral considerations—both strict duties and various moral reasons. Because of their importance, we may not want to simply apply a rule and move on. We may want to know whether there is, in fact, a uniquely right answer to what we should do. Does our participation in collective harms and benefits never decisively rule in or out some action? And how do these considerations interact with more traditional duties?

It is to these questions that we turn now.

CHAPTER 13

Monumental Ethics: The Case of Having Kids

> If children were brought into the world by an act of pure
> reason alone, would the human race continue to exist?
> Would not a man rather have so much sympathy with
> the coming generation as to spare it the burden of exis-
> tence, or at any rate not take it upon himself to impose
> that burden upon it in cold blood?
> —Arthur Schopenhauer, *Studies in Pessimism*

In October of 2010, I found myself attending a series of talks at the
annual meeting of the Virginia Philosophical Association in Ar-
lington. That's where I first met Tina Rulli, who was finishing her
PhD at Yale. The talk she gave to a large room of philosophers was
about the subject of her provocative dissertation, which argued
that those who want to be parents should not create a new child
but should instead adopt an orphan who already exists and needs
parents.

Tina was already, as a graduate student, a great speaker, and I
was captivated by her presentation. I have two stepsiblings who

were adopted out of orphanages when they were young (but no longer babies, so the orphanage was less likely to find a home for them), and I often thought that their mom had done something really special. By choosing to raise children who already existed, she got to grow her family, which is something that she wanted, and the children got a warm, loving home, which they needed. Thinking about all the good that adopting promotes, I have said to many people over the years that I think it must be one of the most praiseworthy actions someone can take.

Tina disagreed with that way of putting it, though. It's not "merely" praiseworthy. It's obligatory—the subject of a strict duty. If you plan to have children, you are morally required, she thinks, to adopt rather than to create a child of your own. While her claim is surprising to many, the argument for it is very strong.

I was intrigued.

Tina's primary argument for the duty to adopt is based on a principle of rescue: if it's relatively easy for you to perform a rescue, then you are obligated to do it. The classic case used to motivate the principle of rescue is "The Drowning Child," which asks you to imagine that, as you are walking by a shallow pond, you notice a young child drowning. There is no one around, the child is in obvious danger, and the only cost to saving them is getting your nice shoes wet when you wade in. The case is designed to elicit the intuition that yes, you are obviously obligated to rescue the drowning child at the minor cost of damaging your nice shoes. The rescue principle then proposes that this is because we have a general obligation to perform rescues, especially when they're easy.

Although any parent will tell you that parenting isn't easy, and so we might think that adopting a child is not an easy rescue to perform, Tina makes the point that potential parents (those who want to parent a child) are already prepared to make the sacrifices constitutive of parenthood. They already know that becoming a parent will sap their time, energy, and money, and are planning to do it anyway, so the general difficulty of parenthood is not a reason not to perform the rescue. Further, while many potential parents argue that adopting would be difficult for them because they desperately want a biological child, Tina shows that the reasons they offer for wanting a biological relationship (they want the child to share the parents' biological features, attitudes, or talents, say) are not guaranteed by the genetic relationship, are trivial desires compared to what is at stake for the potential adoptee, and are often satisfied by adopted children (who can still, of course, share a parent's attitudes and talents by virtue of being raised by them). Wanting your child to "look like you" is not a good reason not to perform a rescue, she argues.

It's an uncomfortable argument to confront, and my interest was not solely academic. I was closing in on thirty years old, Sadiye and I had been married for several years, and we regularly talked about what kind of family we wanted to pursue. We were attracted by the idea of adopting, but I found it difficult to believe we were morally required to do so.

So I kept thinking.

Unfortunately, my reflections were not comforting. The more I thought about the ethics of procreating, the more I recognized an uncomfortable truth: there are lots of moral reasons not to create new people. Tina's argument is powerful and important to consider. But it's not the only one. If we want to be morally serious about having children, we have to face the fact that there are many compelling reasons to think that what most of us believe is wrong.

The first category of arguments against having a child isn't about that child at all; it's about resource use. This is where Tina's argument fits in. On her view, what makes it problematic to have a biological child is that there is a dramatically limited resource in the world, which is the willingness of adults to parent children. The importance of this resource is due to the fact that there are children who already exist without parents, and not enough potential parents willing to raise them. Tina, in other words, is focused on the demanding, taxing, time-consuming will to raise another human being, and pointing out that, as long as there remain orphans in need of parents, those who have that will must decide whether to perform a rescue of one of them or to create another child to spend their parenting resources on.

But there are other resources at stake too. Parenting is very expensive. According to the USDA's Food and Nutrition Service, middle-income, two-parent families in the United States should expect to spend over $233,000 raising each of their children through the age of seventeen, before the cost of a college education is even figured in. Higher-income households are expected to spend more than $372,000 per child. This amounts to putting a huge amount of resources into a single child, when prospective parents could do much more good by using that money to fund other projects—like, say, charitable giving directed at helping the poorest people in the world. So according to Tina's argument, those who plan to parent should adopt rather than procreate. But it could also be the case that each of us ought to forgo parenting in any form in order to do more good with our money.

Procreation, then, is accused of being a bad use of resources—either one's parenting resource or simply one's money. If we are obligated to be good stewards of our resources, then it looks like we could be either obligated not to have children at all or to adopt any children we want. But note that this first category of arguments not to have a child doesn't have anything to do with the child themself—procreation is a problem only because we should really be doing something else with what we have. The second argument against the morality of procreation shifts the focus to what we owe potential children.

Antinatalism—or the view that there's something bad or wrong about having children—has a long history of lamenting what a terrible thing it is to bring a new person into the world. Arthur Schopenhauer famously claimed that life is so full of misery that a sympathetic person would never impose it on another. More recently, the philosopher David Benatar has argued that coming into existence is always a harm, and so for each of us, it would be better never to have been. As a result, he believes that each of us is obligated not to procreate, since never coming into existence is to be preferred over existence.

Even for those who reject these more cynical antinatalist views, I'm sure many parents or potential parents have worried about bringing their children into a scary, fraught world. Some historical moments may make this particularly acute. We can imagine that many parents in the early 1940s wondered whether it was reasonable to start a family during World War II, when the future looked potentially very bleak. Or during the Cold War, as nuclear weapons proliferated and it began to seem inevitable that eventually, someone would make a terrible mistake and press the button that began a chain reaction of mutually assured destruction between superpowers.

Today feels like another of these historical moments when the future can look very bleak indeed. Climate change threatens to make the planet less hospitable to human life over precisely the time span that a child born today can expect to see. Those most worried about the coming damage from climate change may think that, for the first time in some generations, we have good reason to expect that our children will have worse lives than we have—and some reason to fear that our children will face severe hardship (if, say, we don't end up significantly curbing emissions and are on a path to somewhere in the neighborhood of 4 degrees of warming this century). We don't need to be unreasonably pessimistic to worry that the world we're creating may not be kind to any children we create—that the beings whom we will predictably love the most in the world may suffer as a result of our bringing them into existence. And this is a terrible thought. Paradoxical as it may sound, if we have a dim enough view of the world and the immediate future, we may want to protect our children by not having them.

The last argument against procreation shifts the focus from whether the existence of another person would be bad for that person to the question of whether the existence of another person would be bad for the world. Superficially, this case against procreation is rooted in the view that the world is "overpopulated," and so adding a person to it is contributing to a severe problem. But the language of overpopulation is problematic, so we should be careful if we are to give this last worry a fair shake.

The language of overpopulation invites the inference that there is a certain number of people that our planet can host, and we have passed that number. But thinking of that number as some static feature of the world is overly simplistic. Overpopulation is about resource use: an environment is overpopulated when the relevant species is using more resources than it can sustainably provide. But that means that human overpopulation is not just about the number of people on Earth—it's about the number of people on Earth, consuming at a certain rate. And what are they consuming? Well, a whole variety of limited resources, including obvious ones like land and fresh water, but also less obvious ones, like the atmosphere's ability to absorb greenhouse gases without violently disrupting the climate.

So what exactly is overpopulation? An analogy may help. Think about the Earth's natural resources as our collective inheritance, which we can invest by simply leaving them alone. So fresh water, clean air, the rainforest, swamps—all of it makes up the wealth that will take care of us, and any part of it that we leave alone will generate interest. In fact, the planet gives us an incredible return when we're responsible, spinning off a huge amount of natural resources, sufficient to feed, clothe, and generally take care of billions of us. If we use only what the Earth can replenish each year, then it can provide for us indefinitely. In such a case, we would be living off the interest of our eco-savings account, leaving the capital untouched. That's what it means to live sustainably, because we can spend resources for our entire lives and yet leave just as much for our children as we had inherited from our ancestors.

The question of whether the planet is overpopulated is whether we have created too large a population consuming too much, such that we've passed the point of living off the interest and begun eating into the Earth's capital—whether we have passed what scientists call the Earth's "carrying capacity." And unfortunately, there is strong evidence that the answer is yes. The Global Footprint Network, for instance, attempts to determine at what point each year humans have used all of the biological resources that the Earth can regenerate in an entire year. In 2022, that date was July 28, meaning that for nearly half the year, we were in "overshoot"—using more resources than the Earth can provide sustainably. We're spending down the capital. Other scientists have tried to calculate how big a sustainable human population would be. Obviously, that depends a lot on details of resource use in that population, but some estimates have been worryingly low—in the neighborhood of 2 to 3 billion.

People often have a hard time accepting that the world might actually be overpopulated. There seems to be so much space. Visitors to the American West marvel at the hundreds of miles of nature with so few people in it. But the idea of the Earth's capital is supposed to make clear how our planet can be overpopulated even when there are large plots of land where we could fit yet more humans. Those humans would need resources, and if we spend more than our annual ecological interest, we eventually face bankruptcy.

This final argument against procreation, then, points to the evidence that the Earth is overpopulated, and so suggests that there is something morally wrong with adding yet another person to the problem. Yes, the world might be bad for your kid; but your kid might also be bad for the world. In both directions, then, there is moral risk to creating a new person. And that's all in addition to our starting point that raising a child comes with severe opportunity costs for doing moral good.

The case against having children is distressingly strong.

As I'm writing this, my (biological) daughter is downstairs practicing violin, so I clearly decided it is permissible to procreate (either that or I just came to terms with doing the wrong thing). But I did—and do—find the question genuinely difficult. The case of having children is a paradigm instance of monumental ethics in which classical ethical worries combine with the moral pressures from threats like climate change to make ethical deliberation truly difficult. Let's look at the arguments against procreating to see if, when, and how they might be defeated, and what that reasoning looks like.

My response to Tina's adoption argument is the most straightforward, because we have already reviewed the moral tools needed. The question is not whether we are sometimes obligated to rescue—I agree with her that we are. In the case of the drowning child, you obviously have a duty to wade into the water and rescue the child, even if it means ruining your shoes. When we apply the principle to adoption, however, we end up with an incredibly invasive claim: you are obligated to adopt rather than procreate, which is to say that you do not have the moral freedom to decide how to form your own family. My invocation of moral freedom, however, should tip you off that I think something has gone awry. Deciding how to form a family seems like the kind of thing that we should want to have extensive moral latitude concerning.

Here, I'm leaning back on Maggie Little's argument from the abortion case that some actions are so intimate that they are not typically the appropriate target of a positive obligation. In that discussion, we reviewed many sorts of actions that are intimate under some description: sex, marriage, kidney donation, and of course, gestation. My addition here is that bringing a child into your family is relationally intimate in a way similar to marriage. You are choosing who to form a life with, who to bind yourself to forever, and that's not the sort of thing that tends to be the subject of a duty. Just as you are not obligated to marry any particular individual, even if it would be great for everyone, you're not obligated to have a child through adoption, even if it would be great for everyone.

Before we are too comforted by this response, though, we should remember the second half of Maggie's view: just because you don't have an obligation doesn't mean morality is silent on the issue. Having a child in one way over another can reflect badly on your character, and you might have most reason to do something other than you do. Suppose I want to have a biological child because I really care about passing on my genes. It seems to me to be important that a child of mine isn't merely raised by me but is *of* me in some deep biological sense. Given the needs of existing children, if I chose to have a biological child rather than adopt solely for this reason, you might have serious questions for me. Like: Why do I think my genes are so special? What exactly am I accomplishing by passing them down? If I don't have good answers to these questions—if I just think I'm from especially good stock—then you might think my choice reflects badly on me. The idea that my genes are especially good might seem problematic in many ways, both a bit narcissistic and a little too close to eugenic ideologies (that it's important to pass down "good genes" from "the right people"). And since that doesn't seem like a good reason to procreate, choosing not to adopt a child who already exists and needs parenting resources can look selfish. That still doesn't make adopting obligatory, but it makes procreating for that reason morally problematic.

The argument for giving to charity instead of procreating is broadly utilitarian in that the motivation here is to do more good with one's resources rather than less, even if that means forgoing something of great personal value (like being a parent). Thus, my responses should be unsurprising given our exploration in earlier chapters.

The first problem with this view is that many of us take at least some degree of charitable giving as not obligatory but rather as *supererogatory*—above and beyond the call of duty. While it is worth debating whether some amount of generosity is, in fact, required by morality, most of us think that giving to the point of genuine, true sacrifice is good and praiseworthy but not required. Given that people who want to be parents would see remaining childless as a genuine sacrifice, it is difficult to believe that refraining from procreating so as to promote more good elsewhere in the world is obligatory. This is an instance of the broad "demandingness" objection to utilitarianism.

We should also revisit Bernard Williams's integrity critique, noting that an obligation to promote good at all costs turns us each into happiness-promoting machines. To have integrity is to have valuable life projects that we are allowed to pursue, even when they are suboptimal from the perspective of happiness-promotion. Yes, it is likely that the money I spend on my daughter could produce more happiness if I donated it to a foundation that redistributes to the poorest people in the world. But if one of my life projects is to be a father, then a moral rule that prohibits me from doing that because I could do more good elsewhere also prevents me from living out my values.

Of the arguments against procreation, I must admit that the charity argument has the weakest grip on me. But even here, we should acknowledge that the point made is not nothing. The argument is still a reminder of the moral opportunity cost of procreating. By having a child, I commit to channeling a huge number of resources into a single person, despite the fact that there are millions of already-existing people who need my resources. That does seem to be a moral cost, and so in our final reckoning we should remember to ask, "What does my choice to have a child mean for how I live a decent life moving forward?"

The cynical, Schopenhauer-style argument is the darkest to consider. Most of us don't want to imagine that coming into existence could actually be a harm, or that our children may come to hate us for creating them. And indeed, most of them won't (however much they may protest during those rough teenage years). Which is part of why Benatar's argument seems unbelievable to so many of us when we encounter it. If people largely enjoy existing, or at least prefer it to not existing, on what grounds should we say that coming into existence is generally a harm?

Benatar thinks that the answer is twofold. First, people are largely deluded about how bad their lives are (this is an important psychological coping mechanism, he thinks), as an objective evaluation reveals most lives are full of discomfort, some modest suffering, some severe suffering, and eventually death. Second, we typically compare our lives as we know them now with the thought of going out of existence to justify the view that existence is preferable. But this isn't the right comparison, since he's asking whether we should ever bring someone into existence in the first place. The right comparison is whether our lives (full of discomfort and suffering) are better than never coming into existence at all, in which case we wouldn't miss the good things we happen to like now that we exist.

My primary objection to Benatar is that I don't think you can be harmed by being brought into existence. Not just that we typically aren't—I don't think it's conceptually possible. And that's because harm is a comparative notion. If you punch me in the nose, I'm harmed because I went from not having a broken nose to having one. You made me worse off. The thing about coming into existence is that it can't make you worse off. Why? Because you didn't exist before, well, coming into existence. Nonexistence isn't some property that we can have, like, "I used to be nonexistent, but now I exist." That's a confusion. Before coming into existence, you didn't have properties. There was no you.

This is all very weird, I know. But it's important! We're talking about bringing people into existence, which is an activity that we humans are big fans of. So you can't hurt someone by creating them, and for similar reasons, you can't benefit them either. Some people like to talk about the goodness of existing, as if it's a gift we can bestow on someone. But that's just as much a confusion as thinking we can harm them. You can't bestow the gift of existence onto someone because there is no one there to receive that gift. Creating someone is the act that makes it possible for them to ever be benefited or harmed.

Some philosophers do think there's one exception to this: you can be made worse off by being created if your life is so bad that it's not worth living. What does that look like? It's not exactly clear what would make a life not worth living, but we can imagine at least some cases. Imagine a potential parent who has a genetic condition such that, if they were to procreate, the resulting child would have devastating biological anomalies. If, in this case, the child would be born into a world of unremitting suffering, with no prospect of healing or development, followed by death after only a few months or years, we might think that's a clear case of a life not worth living—all bad, no good. And so some philosophers say that such a child would be harmed if they were made to exist with that condition, because their life is worse than never existing. To be honest, I'm not sure this makes real sense, since it's still the case that no individual is made worse off, so it feels a bit like a cheat. It does seem wrong to create such a child, but that wrongness doesn't seem to depend on the idea of harming the child. In any event, the discussion of "lives not worth living" is fairly immaterial for our purposes, since the cases we're considering are the much more mundane ones of normal, run-of-the-mill suffering that comes with existing in the world.

We should not be overly comforted by the idea that no one is harmed by being created, however. Because, as is my wont, I see much that still matters in the cynic's argument. It still seems to matter morally how well I think a potential child's life will go. Consider the following case: Imagine that you and your partner are trying to have a child, and you come down with an illness. From your doctor, you learn that your illness will resolve in a month, but if you create a child while ill, it will live a life full of disease, surgery, and pain. However, if you simply wait a month until your illness resolves, any child you create will avoid that fate. Should you wait a month to procreate?

Obviously you should, right? You should wait, because it is a serious moral cost to create a child who will live with pain and suffering when you could instead create a child who will not. It's not that you would harm your child by creating them, but there is something wrong with intentionally choosing to create a child who will have a worse life. The potential suffering of a future child is a reason not to create them. Now, contra Benatar, we might be comfortable saying that if the suffering is minor, that's not a very strong reason not to create them. Knowing only that my child will have the average day-to-day mix of unhappiness and happiness might not seem like a particularly strong reason not to procreate. But if I could instead have a child who will suffer less? Should I do that instead? If we think the answer is yes, then it seems the suffering matters some, even if it's not decisive.

These are the sorts of worries that occupy me when I think about my child growing up in a world that may include catastrophic climate change. I wonder if I should have protected her from that world by not creating her. Of course, if I had, she would not have experienced the joy of play, the warmth of a family's love, and a million other good things. And so many people will encourage me to balance the harms she experiences with the benefits of her life, and to optimistically predict that the benefits will still outweigh the harms. But here's the last strange bit of philosophy concerning coming into existence: while potential harms seem to count against creating the person who will experience them, potential benefits do not seem to count in favor of creating the person who will experience them.

This very strange claim is called the *procreative asymmetry*, and it is justified by intuitions like those I've already outlined (you should not create children who will predictably suffer greatly) combined with the following question: If you happen to be a fertile person, does the fact that you could, right now, create a person who would experience joy give you any reason whatsoever to create them? I could be adding more people to existence, and I can reasonably expect that they would live fairly happy lives. But that expectation does not seem to provide any moral pressure to get to work making babies. It's not just that I seem to lack good reason to make more babies right now. I seem to lack *any* reason. So the risk of my child having a bad life counts against creating them, while the promise of their having a good life does not count in favor of creating them.

Some folks have objected to me that the procreative asymmetry is so conceptually strange that we should just accept that we have reasons to create happy people. After all, those reasons might be outweighed by other considerations, like my ability to manage my life with a certain number of children (or by finances, or whatever). The problem is that promoting happiness is not just any old reason; it's a moral reason. And when moral reasons add up, they exert real moral pressure. In my life, with my resources, I have every reason to believe I could raise several more kids with a reasonable shot at being happy. If there was significant moral weight favoring making happy babies, then I should experience not having more kids as a moral loss. If I understood morality's call, and if I were stronger-willed, I should have done it and it would have been good. And if I end up choosing not to procreate because I don't want to, because kids are hard work and exhausting, then I should seem selfish, because those aren't good reasons not to promote happiness. But that doesn't seem to be the moral situation. When I chose not to have any more kids, regardless of my reasons, that seems totally fine. There is zero moral pressure to be cranking out happy babies.

What seems true here is that there is a crucial moral difference between creating happiness by making an existing person better off (we always have reason to do this and sometimes an obligation to do so) and creating happiness by creating someone to experience that happiness (we have no reason or obligation to do so). In a slogan: we should make people happy, not make happy people.

Considerations of our children's interests weigh at least somewhat against procreation, which is depressingly close to admitting that Schopenhauer was onto something.

To this point, I have offered a qualified defense against antinatalism. There is genuine moral pressure both to adopt children who need parents and to use our financial resources to promote the well-being of the worst off, rather than creating someone new to pour those resources into. But we are not obligated either to remain childless or to adopt. Others may not demand this of us, though each instance of procreating is morally serious. In addition, the world into which my daughter was born is scary, for sure, but she was not harmed by being created. So I had no strict duty to prevent her from experiencing both the suffering and the joy of existence.

This defense is qualified, because for each argument, I conceded the existence of moral reasons that compete with procreation. Not everyone will have a good justification for choosing to have a child. We need, then, to further explore any reasons one may have for or against growing their family in order to comprehensively evaluate the case for procreation.

Which brings us to the final piece of the case against procreation: the overpopulation argument. In considering the role of the planet's resources in the decision to procreate, we find ourselves back in the world of participatory ethics. If the arguments of this book are broadly on target, the presence of 8 billion other people on the planet does not generate an obligation not to create another mouth to feed. Even if we grant that the Earth is in overshoot, and so we are eating into the planet's capital, that does not mean that my having a child meaningfully worsens overpopulation. But it also doesn't feel appropriate to simply apply one of our heuristics to the monumental decision of whether or not to have a child and then move on. For instance, one of my climate-related heuristics is "Don't be wasteful," but even if we recognize that creating a new person uses resources, it isn't wasteful. People have children because forming a family can be wonderful and a source of great meaning. It's not like taking a seventeen-minute flight on a private jet.

What makes the ethics of procreating against a backdrop of limited resources so difficult is that having children seems both uniquely valuable and uniquely costly. While the value side is easy for most people to appreciate, we should look more closely at the costs.

Most of our private, individual actions have a limited scope of environmental impact. When I purchase something, or travel, or whatever, there is a calculable one-time environmental cost. But this is not how procreation works. Yes, having a child increases a whole bunch of one-time environmental costs (you buy diapers, toys, a bigger car, a bigger house, and so forth). But much more important than that is the fact that in creating a new person, you create another being who will have their own environmental footprint—and may then create more people who will have their own as well.

Although overpopulation is a problem for many environmental reasons, arguably climate change is the most acute and severe, since, as we saw in Part I, curbing emissions now is of paramount importance. So we can use greenhouse gas emissions as an example of the environmental cost of procreation.

The fact that future generations will have their own carbon

footprint is the problem of carbon legacy. In one study, scientists tried to calculate the carbon legacy of procreation by assigning procreators a dwindling percentage of their offspring's future emissions. In order to do this, they used the intuitive idea that each of us should take ourselves to be responsible for an amount of our descendants' emissions roughly equal to our genetic contribution to them. The formula, then, is that I am responsible for $(\frac{1}{2})^n$ of my offspring's emissions, where n is the number of generations they are removed from me. So my partner and I are each responsible for one-half of our child's emissions, one-quarter of our grandchildren's, one-eighth of our great-grandchildren's, and so on. Now, they might be wrong that this is the right way to think about responsibility for one's offspring's emissions. One could reasonably think that procreators are responsible for more than half of their child's emissions, since each of us who creates a new person makes all of their emissions possible. But since children become agents of their own, procreator responsibility dilutes more quickly through each generation.

Regardless of the exact model we use for determining one's responsibility for their offspring's emissions, carbon legacy makes it clear that the choice to have a child is among the most consequential decisions many of us will make. In procreating, I'm making a new consumer and emitter—and maybe a new procreator. I'm starting a train of emissions, which will continue so long as humans are net-positive emitters (which is hopefully only for one or two more generations, but of course we don't know this).

Given the severity of the climate-change threat, actions that have a larger carbon impact seem to be better targets for reduction or elimination, which is why the reasons not to fly in private jets (or at all) are stronger than the reasons to compost your lunch leftovers rather than throw them in the landfill. Both reasons are real, and neither amounts to a duty, but they vary in strength. You need a better reason to justify flying than you do to justify not composting.

If we take that reasoning to its logical conclusion, then, the decision to have a child gets put under serious moral scrutiny. Most of us have stronger reason to eliminate or reduce our procreative activities than we do to change practically any other behavior, and so we need stronger reasons to justify our family choices than we do in other domains.

And now I think we're finally in a position to fully evaluate the ethics of having children in the twenty-first century. Although my partner and I had many reasons for creating our family to be the shape that it is today, I often think that climate change is a major reason why my daughter is an only child. But that idea has two parts that are worth investigating: having one child implies that I think procreating is justifiable; but having only one suggests that I think having a larger family wasn't (for me).

The reasons that count against procreating are serious. Although I don't think you harm someone by creating children, existence comes with massive risk, and we happen to live in an era where that risk is heightened. Creation also brings with it other moral costs—both the relatively large contribution to environmental problems, and the opportunity costs of not pursuing other worthy projects with either one's parenting resource or one's financial resources. What all of this means, I think, is that those of us who choose to have children owe justification for doing so in the face of real reasons to refrain. Those reasons call on us to stand up to them and offer a compelling case for becoming biological parents.

Some reasons that have been offered to me sound reasonable at first blush but don't stand up under scrutiny. One of the most common of these is the idea that my kid may be the one to save the planet from problems like climate change. So I have good reason to procreate, because more people means more ideas and more chances to solve hard problems. But it's difficult to put too much weight on such a consideration. Almost 400,000 babies are born each day, which makes it difficult to believe that my kid is so important to the future of humanity. Humanity is already buying lots of lottery tickets; it takes some serious hubris to think that mine would make the difference. And of course, while it's possible that each child will be the next Einstein, it's also possible that each child will be the next serial killer or cult leader. It's a dark thought, but if we're relying on fortune and probabilities to justify needing another child on this Earth, we have to be honest that fortune and probability cut both ways.

Of course, some people who raise this point might be thinking that it's not mere fortune that determines how likely it is that a child will be important. They might be thinking that "good people" are precisely the ones who should be having lots of babies. But now we have another worry, as humans have tried this sort of reasoning before and it didn't work out too well. It's called eugenics, and it's tightly tied to racist and classist ideologies about what kind of people it's good to have. There are no morally defensible distinctions between "types" of people such that we should get into the business of trying to figure out who should be having babies in order to save humanity.

One need not think that any baby would be special, however, to think that it would be good for the world to have another person in it. And this is where things get interesting. Just as I worry that another child is another consumer and emitter, some point out that each new child is a consumer and producer—that is, they are another productive member of society. They are another tax-payer, another person to take care of the elderly, and another cog in the capitalist machinery, creating a tiny bit of GDP. This sort of reasoning is not new, as there is a long history of nationalist expectations of families (read: women) that they create new members of society to fulfill important needs, including providing soldiers for the battlefield. In all these cases, the basic idea is that society needs people, and so by procreating, you are providing a tiny little piece of society. This concern has taken on greater urgency in recent years as countries like Japan have seen their populations shrink and their demographic pyramid begin to invert, with more and more people getting older but fewer young people to run the economy and care for them.

On the one hand, this argument makes perfect sense: if we can identify collective goods, like the functioning of society, then there is value in participating in those goods. So we can identify participatory reasons to have children in order to grow one's society. However, these reasons are suspect on moral grounds in at least a couple of ways. First, the economic argument is based on the idea that economies (and so populations of consumers and producers) must grow forever or else they'll die. But the issue of overpopulation is precisely the problem that infinite growth in a world of finite resources is impossible. Thus it's not clear that there is value to growing a population past sustainability. And second, there is something morally worrisome about the idea that one's reason to have a child is for some other good. It treats the child as a means to some larger moral project or goal, rather than as a person who should be respected. This is another moral asymmetry when it comes to procreation: the bad that procreating contributes to seems like a perfectly good reason not to have a child, whereas the good that procreating contributes to seems like a morally suspect reason to have a child. In the negative case, there exists no child who was treated as a means to the end of contributing to some global project. But in the positive case, there does exist such a child, and it seems disrespectful to have created them for the sake of promoting a project that has nothing to do with the child itself.

Thus, I think that many of the commonly offered reasons in favor of having a child don't justify it. However, that doesn't mean that there aren't good reasons to be found. Becoming a parent is among the most valuable life projects that some of us envision for ourselves. And the value of this project is deeply connected to the child brought into the world, as they get to become part of something special—namely, a family. But becoming a parent does not distinguish between procreating and adopting. So why have a biological child over adopting? I think Tina is right that many of the reasons offered ("because I want my child to look like me") don't stand up to the moral pressure or reasons to rescue an existing child who needs parents. But there are good reasons to procreate for some people, because creation is a genuinely special activity. Sadiye taught me this long ago, when I confided in her that I didn't have a strong enough interest in procreating, and so perhaps we should adopt. Her response was that creating another human with her body—growing it, feeding it, making it out of her own flesh— was among the central goals of her life. She didn't only want to become a mother; she wanted to create. She described the pull of a valuable life project that I hadn't experienced, since I can't gestate. But once she described it to me, I could clearly see its force.

Parenting is a valuable life project, but for some people, creation is too. Such a desire to create certainly isn't sufficient to make procreation permissible—wanting to experience gestation without wanting to be a decent parent would not justify creating a new person. But for the prospective parents who want to create something of value by building a loving family, this drive to gestate seems like precisely the right sort of reason to stand up to the demand for justification.

So I don't think it's difficult to justify the decision to have a child, but it's important to do so, because not everyone who procreates actually shares these values. Some parents don't want to gestate and deliver a child. Some people find the thought terrifying, considering what a serious medical risk pregnancy carries. Some are simply neutral about the prospect—gestation is not taken to be a central life project, but it is something they're willing to do to become parents. And for these folks, choosing to procreate anyway may be unjustifiable. There are powerful moral reasons to do something else, and they don't have a particularly strong response to those reasons. The requirement to justify one's actions isn't important because the justification is necessarily difficult to meet; it's important because it reveals that, even when a justification is available, not everyone has it.

Additionally, those who do have central life projects around procreation may not thereby have good reason to procreate repeatedly. When we had our daughter, Sadiye got the opportunity to create life, and we became parents. The value of these transitions in our lives justified what we accept as the serious moral cost of having a child. But the next question is whether we would be justified in doing it again. Having one child fundamentally changes who you are, and gestating and birthing makes you a creator of human life. These are incredible experiences and important changes to one's identity. But when we considered having another child, we didn't feel as strongly about it. Neither of us attached significant value to having a large family, and Sadiye was satisfied that she got to create life once. We no longer felt able to stand up to the reasons against procreating.

Our first decision, then, was that we would adopt any other children after our first. But adopting wouldn't mitigate the worry that children are expensive, which creates a moral opportunity cost of having more children. We have strong commitments to our families, loved ones, and various other moral projects, and so we felt we would need to really value having a larger family to justify pouring resources into another individual. And it turned out, we don't.

This sort of deliberation is particular—it was specific to us as a couple, and it would be different for anyone else. No one has the standing to demand that you have a family of a certain size, because we are operating outside of the realm of duty and obligation here. Rather, we have a variety of reasons not to procreate, to adopt instead, to promote the happiness of already-existing individuals, and to not participate in the harms of overpopulation. These reasons call on us to justify our decisions, but we can, in fact, do so. In the realm of the intimate, our values and the goals around which we structure our lives carry a lot of weight. Taking procreation as a morally serious decision worthy of real scrutiny could lead to more people deciding that remaining childless is the right decision for them. But it's also possible that some people, for whom having a large family is a life-organizing project, can justify having many children—even on a planet where resources are scarce and getting scarcer.

Which leads to my last point, which is that justifying a decision does not make the opposing moral considerations go away. My own choice to have a child makes me a large consumer of precious resources, especially because my daughter was born into relative privilege. The reasons not to take more than one's share generate a moral remainder when we act against them, even when such action is justifiable. Getting to have my daughter, who is an incredible joy to me but expensive for the planet, puts me on the hook for paying back into the environment. I have yet more reason to fight climate change in all aspects of my life than I already did before having a child. The same is true for all the other reasons I chose against in deciding to have a child. I put her at risk from existence, which puts me on the hook for doing everything in my power to mitigate that risk. I am overinvesting my resources into just one child, so I should be on the lookout not to do so thoughtlessly, and to be generous with those resources (both parenting and financial) when it comes to others.

The various challenges from antinatalism often get dismissed without much investigation because they make us uncomfortable. We might think that no one should be able to comment on the morality of such a deeply personal aspect of our lives. And that may be true—it is no one else's place to determine the justifiability of how you formed or will form your family, and certainly not their place to speak of it without an invitation. Here, I've said only what the arguments about procreation have meant for me; I can't say what they imply for you. But even if it would be rude or inappropriate for others to comment on the morality of procreation for anyone other than themselves (without invitation, anyway), that doesn't mean morality doesn't operate in that space. Moral reasons exist whether anyone points them out or not.

Choosing whether or not to bring a new person into existence is among the most monumental ethical decisions that many of us will ever make. Living a good, decent, conscientious life requires that we make a good-faith effort to respond to the plethora of relevant moral considerations. Doing so is the only way out of the Puzzle that children represent.

CHAPTER 14

Being a Moral Participant

> In the absence of concrete economic and legislative changes, consciousness raising through anti-racist reading is mere filibustering—white people learning about their privilege and power without ever having to sacrifice either.
>
> —Saida Grundy, "The False Promise of Anti-Racism Books"

In the summer of 2020, during the awakening of (some) white people after the killings of Ahmaud Arbery, Breonna Taylor, and George Floyd, protests erupted in cities all around the world, forcing those in power to have a long-overdue reckoning with continued forms of racism. Many communities had some of the most sustained public discussions in recent memory about how white people can do better at understanding and promoting antiracism. Reading groups popped up to discuss Ibram X. Kendi's *How to Be an Antiracist* or Ijeoma Oluo's *So You Want to Talk About Race?*

Concepts like white supremacy and white fragility came up in ever more conversations.

Although I imagine that most antiracists saw this sudden drive among white folks to educate themselves as a positive development, there was eventually a mild backlash. Are more book clubs really what we need in the face of rampant, violent racism? Might the push for individuals to educate themselves actually be a way to undermine more concrete, structural changes? Tre Johnson made the criticism powerfully, writing in *The Washington Post* that his white friends have learned the right "pieties" to perform after racial killings, consuming media by people of color and sharing the profound insights that they've thereby learned. And so, "When things get real—really murderous, really tragic, really violent or aggressive—my white, liberal, educated friends already know what to do. What they do is read. And talk about their reading. What they do is listen. And talk about how they listened. What they do is never enough."

Johnson's concern was echoed in think pieces and on social media, as activists and scholars worried that well-meaning white folks may read some books or articles, listen to some podcasts and do some soul-searching, and then be awfully glad that they got that whole racism thing figured out.

In other words: racism, despite being a very different kind of catastrophe, has some of the same properties as other structural threats that give rise to the Puzzle. While many (especially white) people think of racism as an attitude—as hatred, suspicion, distrust—that is only one of the forms that racism takes. And ethically, it's the easiest one to address, because it wears its ugliness on its face. But racism is also about policy and structure—about a status quo in which systemic forces make it easier for white people and harder for people of color to achieve the same thing. Structural racism is about a pay gap between whites and racial minorities; about differences in life expectancy and healthcare access; about lack of representation in professions, politics, and virtually every other sphere of public life; about the additional work that people of color have to do to even have a chance to achieve or obtain some of the things that white folks are often born into. And of course it's about a system of policing and criminalization that leads to violence against Black people in particular, often in situations where a white person would never have to fear for their life.

One form of the backlash against antiracism book clubs, then, was the claim that white people working on their virtue by reading, learning about other cultures, and so on was not going to fix housing discrimination or inequitable loan distribution. Nice white folks trying to root out implicit hate in themselves was not going to make American drug policy less harmful for people of color. Joining a book club isn't bad. But when it starts to be touted as the thing you must do, you'd be forgiven for thinking you've read from this playbook before. Telling individuals that their private actions are going to fix some problem, when entire systems need to be revamped or torn down, is the sort of distraction that Michael Mann warned us about when it came to climate change. Powerful political and legal institutions may well love it when each of us is examining our "racism footprint" if it keeps us from looking at theirs.

This break between individual and collective action reveals that being an antiracist likely has a participatory element. Because racism is not climate change, we shouldn't expect the two cases to be identical. In particular, it's crucial to point out that individuals clearly have strict duties not to act in racist ways, as individual action can cause harm, be disrespectful, and can certainly be an expression of vice. But there are other private actions that may not significantly affect any individual or play a meaningful causal role in the maintaining or dismantling of racist structures, and yet they seem to have participatory value. These are the book club–adjacent activities.

White people, upon realizing that they don't know enough about race and racism, absolutely should educate themselves. In addition, they should think about whether the structures of their upbringing have led them to solely or disproportionately promote the work of white people, through their professional lives and even through the goods and entertainment they consume. Every time I choose a movie to watch, I make a small decision about who and what sorts of industries I support. Because so much of Hollywood has been run by white men for so long, much of entertainment caters to the interests and self-image of white men. Choosing to support filmmakers of color (and you can see how the argument works also extended to women and LGBTQ+ folks) sends a minuscule signal to the film industry and potentially serves a small educational role by consuming entertainment that wasn't created to make people who look like me feel good about themselves. By choosing more diverse entertainment, by educating ourselves on race and racism, by thinking about who we support with various purchases (think of the push to "shop Black-owned businesses"), we can participate in antiracist efforts, which has clear value.

But since such actions will not solve racism, we must consider how else we might participate in the broader fight against racism. Remember that participatory ethics is not just about withdrawing one's support from various structures or collective actions. It's also about engaging socially and politically as an individual. So the warning to book clubbers not to let their reading assuage their guilt and make them feel like they've done their part is crucial. Getting better at consuming diverse entertainment will not solve structural racism, so we need individuals to participate in broader antiracist efforts as well.

Although I have argued that individuals have broad permission to choose which moral fights to participate in, the presence of deep, structural racism teaches us one final, important lesson about triaging our efforts. In the case of climate change, I argued that the acuteness and time-sensitivity made it difficult to justify not including it in one's moral causes. Although a true champion of some other cause may justifiably not think much about climate change, most of us should be developing at least some heuristics for responding to the reasons to fight such a looming catastrophe. Racism is also an acute need, but it has another feature that is important to highlight: its structures benefit white people, whether they want those benefits or not, and so white individuals in particular have an especially strong reason to respond to the presence of racism, and in particular ways.

The argument here—which I made in a paper with my friend and colleague Nabina Liebow—is that, when an injustice benefits you, that benefit brings with it a responsibility to compensate or pay back the victims of the injustice. Since I, by virtue of being a white man, have benefited my entire life by the way people respond to my appearance, I owe my participation in antiracism efforts as a matter of what we can think of as participatory justice. Whereas there are other privileges many of us enjoy that we can "opt out of" to some degree (I can try to opt out of the harms of consumerism by boycotting companies that utilize morally bad practices, for instance), I can't opt out of my race or the way society has racialized me, and so I can't avoid my responsibility by trying to dodge white privilege.

What the racism case teaches us is that the broad latitude we typically have regarding how to respond to structural moral problems isn't complete. There are some cases, it seems, where purity is just not an option. Racism looks to be such a case for white people, because we can't avoid benefiting from how we look just by focusing on ourselves. No amount of withdrawing participation from the systems of white supremacy will make it the case that I don't enjoy unearned privileges. And that means that focusing solely on the top row of the 3 × 3 matrix is never a satisfying option for responding to structural racism. Social and structural activism are an especially important part of responding to injustice when nonparticipation can't be achieved.

There are thus two tightly connected lessons from recognizing that racism raises a version of the Puzzle. We feel the moral pull of responding to a catastrophe in some way when we are benefiting from participation in that catastrophe. Upon such a realization, we can then respond either by withdrawing our participation (focusing on purity reasons) or by participating in efforts to fix the problematic structure. But since structural racism benefits white people without requiring any action on their part, we can conclude: first, all white people all the time are implicated in structural racism, and so they have a special reason to include it among their moral projects; and second, they cannot discharge their responsibility by focusing only on purity. The fact that participation in an unjust structure is the default for white people reduces the moral latitude both in terms of choosing a moral project to focus on and in terms of how to respond to that problem.

Although this structure is easiest to see in the case of racism, I think it is essentially the same for climate change. Climate change is a collective threat that not only causes harm but is a structural injustice in which one group (the world's wealthy) benefits through extractive processes while another group (the world's most vulnerable) suffers the costs. And the benefit enjoyed by individual members of the first group is enjoyed by default, through everyday living. While it is the case that individual emitting actions can sometimes be reduced or eliminated, the structural feature I've referenced throughout this investigation is always there: as an American professor, I can't really withdraw my participation from the injustice of climate change. There's a sense in which it's physically possible for me to repudiate my climate privilege (I could, for instance, try to immigrate to the Maldives or Bangladesh, removing both the degree to which I'm benefited by climate change and increasing the degree to which I'm harmed by it). But this isn't a practical possibility. Living anything like my life, as I've developed it, will implicate me in a massive injustice.

And now I think we're in a position to fully understand the force of the objection that focusing on one's purity is in some way not enough when it comes to climate change—an objection that started way back with our introduction of Wallace-Wells, Heglar, and Mann. We can't actually divest from our climate privilege, which makes focusing on an effort to do so seem performative and unsatisfying. Minimizing one's carbon footprint clearly isn't bad—there are reasons to do it!—but even radical action in that direction won't make it the case that the world's wealthy and powerful aren't benefiting tremendously from the extractive practices that are devastating the planet. So the answer to that endemic privilege is to do more. Withdrawing participation isn't enough, so here, too, we have special reason to develop rules for engaging in social and structural work. For those of us with climate privilege, it's especially important to spend our time and effort trying to tear down the systems that granted us that privilege in the first place.

In Part I, I argued that the Puzzle was far more common than we may initially think it is. Climate change is the paradigm case, but the same push toward what I have called *participatory ethics* arises when it comes to eating, charitable giving, our consumer behavior in general, and infectious disease outbreaks like Covid-19. I've also mentioned the case of voting in massive democratic elections, and now we've discussed structural racism. One of the generalizable lessons from the case of racism is that any problem that is properly described as structural, but about which we tend to moralize for individuals, is likely to have a participatory element, which is to say that there will be reasons for each of us to be part of the solution rather than the problem. And this means that many public-health threats fit the mold of Catastrophe Ethics.

In the United States, more than 100,000 people are dying every year from drug overdose. This is an absolutely tragic figure, representing a catastrophic failure of public health. We could prevent very many of those deaths with better addiction treatment, widespread adoption of evidence-based harm-reduction policies, and changing some of the stigmatizing and punitive drug laws that prevent people from seeking help. And yet we do not, and so year after year, more people die needlessly.

As part of my job, I do a lot of public speaking on this topic, both to the healthcare community and to lay audiences, and I commonly get asked that frustrating question: "So what can I do to make a difference?" It was hearing versions of that question over and over that made me realize that public-health policy failures are a lot like climate change. The harms are vast and based on structures that most of us cannot control, but we also tend to try to rally people into individual action to address them.

So I began to give people a version of my 3 × 3 matrix designed for the overdose crisis, pointing out the different ways in which each of us can participate in collective efforts, from the easiest to the genuinely demanding. When grieving parents who have lost a child to overdose ask me how they can get involved in the fight to make sure no one else has to go through what they are going through, we sit down and talk about what resources, talents, and interests they have to participate in a broader collective effort. Some advice is perfectly general: the stigma associated with drug use and addiction is at the root of the problem, leading people to hide their struggles and resist seeking help when they need it. So what can individuals do? Root out stigmatizing attitudes in themselves (purity); participate in broader conversations that challenge stigma when it arises in their communities (social); and start engaging in local, state, or national politics on drug policy (structural). But some people also have unique abilities. Some have stories to tell that are motivating, and they're willing to tell them. So they write and speak about their experience with addiction or their journey to recovery, thereby organizing their response around their particular circumstances. Some people are doctors or nurses and can put some of their professional energies toward directly helping affected populations. So these folks can become trained to treat addiction and can make their offices a welcoming place for those facing stigma, rallying their education and professional position to participate in the fight against addiction and overdose. And so on. But the basic structure is the same: bad policy, rotten structures, and deep, unquestioned background beliefs in society can lead to catastrophe, and each of us must decide when and how to participate in the fight against them.

An important implication of the realization that participation is a large and crucial part of ethics is that we all must know a lot—all the time, about very many things. You may have already known a bit about America's drug overdose crisis and what we need to do to solve it, or this information may be new. And now you might be wondering what you need to know about the ways in which drug laws and norms are stigmatizing, and how you can respond. If that's you, follow the endnotes to this paragraph, and I'll give you lots of great material to start. But my larger point is that most of us simply don't know enough about all the problematic structures in our world. And if we don't know what's going on, we can't organize our moral projects or develop informed heuristics on their basis.

Return to the consumer examples from earlier. Some of the big questions are more obvious: each of us knows that animal-sourced foods require at least some harm to animals, but do you know just how bad it is? Have you ever read about or seen the torturous practices used in factory-farming systems? If not, there is much to learn before you can understand what is morally at stake in the decision to support animal agriculture. But there are many more ways in which our consumer activities support bad policies or individuals, and not all of them are as obvious. Does the company that makes my new piece of clothing pay fair wages to their workers? Do they engage in sustainable practices? Do they union-bust?

Electric cars are an interesting, complex example. I've noted my support for electric cars because they are a necessary part of decarbonizing society. As long as we're using internal-combustion engines, we cannot fully wean ourselves from fossil fuels, but electric cars can be charged through an electric grid eventually powered by renewable energy. Every realistic plan for decarbonizing sufficiently to avoid catastrophic climate change involves widespread adoption of electric cars, and so purchasing one is a form of participation in the transition to renewable energy.

There are real problems, though. The batteries that power electric cars require rare-earth elements like lithium and cobalt that must be mined, which is obviously not carbon-neutral. And the mining itself can also pose a moral challenge, as rich deposits may be in countries where lax regulations allow for terrible working conditions. The largest cobalt mines, for instance, are found in the Democratic Republic of Congo, where companies have been accused of exploiting so-called artisanal miners, essentially utilizing a form of modern-day slavery. As summarized by Siddharth Kara, "The titanic companies that sell products containing Congolese cobalt are worth trillions, yet the people who dig their cobalt out of the ground eke out a base existence characterized by extreme poverty and immense suffering. They exist at the edge of human life in an environment that is treated like a toxic dumping ground by foreign mining companies." But since the DRC currently supplies about 70 percent of the world's cobalt, many of the batteries found in electric cars likely include materials from one of those mines.

Further, a car company or its CEO may promote values that you find problematic. The CEO of Tesla, Elon Musk, has played a pivotal role in sparking the electric car revolution, but he is not particularly well known for keeping his personal cards close to his chest. Through social media and interviews, Musk makes clear his sometimes contentious personal and political beliefs, which means that purchasing a Tesla can feel like privately supporting this very public figure. Depending on how his values align with your own, that may also seem like a morally salient form of participation.

This single example of purchasing an electric car reveals how difficult it is to know exactly what you are supporting through your participation in various structures and collectives. And it is not unique. Purchasing food and other basic goods, choosing entertainment options, and certainly activities like voting and supporting policies or political candidates all require a lot of information. Which reveals the need for a sort of meta-rule about participatory ethics: we all need to read, listen, and otherwise consume news, analysis, and other forms of information delivery. The moral responsibility to participate ethically in society entails a moral responsibility to be informed. We can't be good climate activists, good consumers, good antiracists, or good participatory citizens if we don't know what's going on in our world.

Interestingly, the very need to be informed generates another possibility of moral participation, as we now live in an era of misinformation and disinformation. While it's clear that individuals have strict moral duties not to engage in disinformation (it's wrong to spread conspiracy theories that may cause harm), the question of participation also asks us to consider boycotting individuals or companies that do so, and to support rigorous, fact-checked journalism. Because we all learn about the world every day, and we often participate in the spread of information through discussion with others, we are all always a microscopic nodule in a vast information network. And like with all of the massive, complex, structural issues explored in this book, we can do our tiny part to make that network better or worse. This involves not only taking care with how we get information, but also taking care not to pass along unverified information and holding others accountable when they do—whether those others are family and friends, or public figures with a much larger podium than most of us have. Participatory ethics in a misinformation age, then, recommends that we all withdraw our support from bad players, financially support unbiased news sources, discuss the moral importance of access to information within our communities, and vote out or advocate against politicians who have far too loose a relationship with the truth.

When I abandoned the idea that private individuals have a strict duty not to joyguzzle, it may have seemed alarming to those with environmentalist tendencies. Indeed, when I began thinking about the Puzzle years ago, I did it with an eye toward beating back the challenge from causal impotence and showing how we can, in fact, justify a duty not to joyguzzle. I thought any other conclusion would be deeply depressing. I was wrong, though. We don't have a duty not to joyguzzle, and that's not a depressing conclusion. Because the lack of a strict duty is not a complete moral license to be an asshole.

The ethics of participation in massive structural and collective actions or outcomes is deeply important and makes up a large part of our moral lives. In an era of multiple catastrophes threatening at all times, we cannot justify inaction, but the old rules of a simpler morality are not enough to advise us. We need new rules, which are unique to us but which are not thereby purely relativistic. The reasons we all have are real, but once we leave the world of demandable, obligatory action, part of determining how to live is determining how I, as an individual, fit into this world. What role do I, with my particular interests, desires, gifts, and handicaps have in the broader collective fight against catastrophe?

When we take that question seriously, the fact that we have no duty to live a particular life should not make us feel like we have been let off the moral hook. We are not left with nihilism. We are left with the rich world of moral participation, which is demanding in its own way. Our moral burden in the face of massive threats is a creative one in which we play an active, agential role in determining the contours of a good life.

That doesn't mean that anything goes. How we respond to catastrophe is the stuff of everyday morality. We all have good reasons to respond to terrible harms and injustices, and a complete failure to respond at all to those reasons is unjustifiable. It's the kind of moral laziness that can make one a bad person, even if they are not violating strict duties. And perhaps even harder to hear: no matter how good we're doing in conscientiously responding to the threats of today, we could always do better. In reflecting on my own personal matrix, I should feel the constant moral pressure to develop and live up to more demanding heuristics, and when I am benefiting from injustice, I should endeavor to do more than the rituals of piousness that focus only on my own purity and instead use my privilege to undermine the structures of privilege. Each of us must do the hard work of triaging the multiple catastrophes of the day, but that doesn't mean that some aren't especially important to address. The acuteness and severity of a threat is clear reason for everyone to address a problem like climate change, but those in a position of privilege have special reasons relative to their participation in the injustice to work to tear down the very system that supports them.

If this sounds too hard, perhaps even overwhelming, let me end by noting what we all gain by facing head-on the challenge in front of us: we gain *meaning*. The threat motivating this book is precisely that we can all feel, in a world of climate change, structural racism, endemic poverty, infectious disease outbreaks, and a million other things, that we don't matter. That what we do doesn't matter. And that way lies the path of moral emptiness. But when we recognize that there is real goodness in responding to threats in different ways, and that we get to—need to—participate in determining that response, we rescue our moral agency from the threat of nihilism. We build a meaningful life.

This is the work of Catastrophe Ethics.

Acknowledgments

This is a big book. Not in the sense of physical heft but in scope. Quite a lot of my core philosophical views made their way into this book, which means that I owe a lot of people thanks.

I mention John Ahrens in the book, who made me want to be a philosopher on my very first day of college. I then found my way to Kate Johnson and Don Carrell while at Hanover, and they corrupted me deeply with their passion for ideas and for being good people. I'm grateful to these folks and to everyone else at my alma mater who had such an impact on my early career. In graduate school, I would be corrupted further, with very many professors inspiring some of the ideas and methods used in this book. My dissertation adviser, Henry Richardson, is an absolute paragon of conceptual clarity, and to the degree that there's any of that in the book, it's due in large part to his mentorship. As Chapter 11 likely made clear, Maggie Little's way of thinking about ethics has probably had more impact than anyone else's on my eventual philosophical development. Although she did not serve on my dissertation committee, she took me under her wing and taught me how to be a philosophically serious ethicist. And for that I'm so thankful.

ACKNOWLEDGMENTS

The seeds of the view I articulate in this book were born many years ago in discussion with my colleagues at the Johns Hopkins Berman Institute of Bioethics. In particular, I learned much in discussion with Ruth Faden, Anne Barnhill, and Justin Bernstein during work on several different projects in food ethics, and Justin would eventually sign on to coauthor my first paper on the topic of the Puzzle. I was then lucky enough to partner with Nabina Liebow, who helped me to think about how the Puzzle does and doesn't apply to the problem of racism.

I hope the book has made it clear how much I owe to the wonderful, inspiring, kind, and generous farmers Liz and Nate Brownlee. In showing me around their land, introducing me to their animals, and speaking with me at length, they helped me to really understand a core component of the view I was working out. They were my philosophical muses.

During the writing of the book, I received tremendously helpful feedback from many friends and colleagues. Marcus Hedahl, Colin Hickey, and Aashna Lal deserve special attention for reading a complete early draft of the manuscript and providing copious detailed notes. Although I'm sure I didn't address all the issues they raised as well as they would have liked, this book is much, much better than it would have been without them. For insightful feedback on particular chapters, I want to thank Maggie Little and Tina Rulli. And for general feedback on the project, I want to thank Kelly Heuer, Anne Jeffrey, Jake Earl, and Tony Ficarrotta.

I also had the opportunity to present early components of this book to several audiences, and I received wonderful questions, comments, challenges, and contributions from each. I am grateful to audience members at the Hastings Center and at California State University, Bakersfield. I want to particularly thank Nancy

Berlinger for the invitation to speak at the former, and Nate Olson for lots of great discussion at the latter. My thanks to Holly Taylor for inviting me to speak at the NIH Department of Bioethics, where I had an incredibly sustained, engaging discussion of the project. While I'm sure I am not capturing everyone, I want to particularly note the contributions of Sarah Conly, Aaron Segal, Kyle Patch, Yukiko Asada, Elika Somani, Dave Wendler, Robert Steel, Dave Wasserman, and George Chingarande. Finally, I had two opportunities to present ideas from this book at my home institution of Johns Hopkins and want to highlight my gratitude for comments and suggestions made by Tali Ziv, Joey Jebari, Em Walsh, Net Lipshitz, Chelsea Modlin, Odia Kane, Maria Merritt, Nancy Kass, and Hilary Bok.

My agent, Jane von Mehren, is simply as good as it gets, and I am so thankful for all she has done to shepherd my work into the world. In this case, that meant finding the very best editor for this particular book—Stephen Morrow—who has been both an insightful collaborator and a champion of the project. Thank you both so very much.

And lastly, to my people: to Sadiye and Sinem, who are affected more than anyone else by my drive to do this strange thing that is writing books. I literally couldn't do it without their support. They also make me *want* to do it. They make me want to fight catastrophe and to leave the world a better place than it might otherwise be.

Thank you all. Whether you agree with what ultimately ended up in the book or not, I want each of you to know that it's better because of you.

Notes

EPIGRAPH

xi   **May I be the:** This line is from "Upstream" in Mary Oliver's collection *Upstream: Selected Essays* (Penguin Press, 2016).

INTRODUCTION

1   **"People who try hard":** This line is spoken by Judge Farris in (King 1978).

2   **cow milk is too environmentally expensive:** It is estimated that producing a glass of cow milk generates three times as much greenhouse gases and uses nine times as much land as any plant-based alternative. See further (Poore and Nemecek 2018).

2   **animal-based foods have:** Supporting data for this claim is widely available, but one way to see it is through the modeling work done by my colleagues in (Mason-D'Croz, et al. 2022).

2   **almond trees require huge amounts:** See further (Fulton, Norton, and Shilling 2019).

2   **all plant-based alternatives:** A good comparison can be read in (Marinova and Bogueva 2020).

3   **with big, mono-crop lawns:** Though not every suburbanite gives in to the ideal of the turf lawn. It turns out that a couple of my fellow Marylanders have rewilded their lawn, and when neighbors tried to force them to clean it up, they sued and won the right to keep a more eco-friendly plot of land (Buckley 2022).

3   **a radically unsustainable cultural choice:** One way to think about the unsustainability of a particular lifestyle is to ask the question: How many resources would we need if everyone on Earth lived like that? This is a question that the Global Footprint Network has asked about every nation's population, and unsurprisingly, Americans' lives are particularly unsustainable. According to their data, if everyone lived like the average

American, we would need 5.1 Earths to supply the resources consumed (Global Footprint Network n.d.).

5 **more than 6,500 migrant workers:** See further (Pattison and McIntyre 2021).

6 **Joshua Greene thinks:** The following paragraphs summarize large parts of Greene's argument (Greene 2014).

6 **Similar to the psychologist:** Kahneman has articulated this broad view in many places over the years, but perhaps most famously in the appropriately named book *Thinking Fast and Slow* (Kahneman 2013).

8 **Humans evolved in:** For most of the 200,000 years that modern humans have been on Earth, they lived in small bands of hunter-gatherers, which tended to range in size from an extended family up to about 150 people. Only with the advent of new agricultural techniques around 10,000 years ago did communities begin to scale up in size, and it was the industrial revolution in the past 200 years that really connected global communities (Aiello 1993).

8 **In the year 1800:** All demographic data in this paragraph is pulled from *Our World in Data* (Roser, et al. 2013).

9 **Greene thinks that:** In particular, he thinks that the dual-process model of cognition is evidence in favor of consequentialism, as it is the moral theory most closely associated with slow, careful deliberation.

CHAPTER 1

13 **"It is worse":** This famous line opens both the essay and the book that Wallace-Wells wrote (Wallace-Wells 2019b).

14 **In fall 2021:** This data was reported in (Associated Press 2021).

15 **During the summer of 2020:** See, for instance, the reporting in (Rosoff 2020).

16 **Iconic beach destinations:** For a thorough look at the situation of the Outer Banks, see (Flavelle 2021).

16 **Miami-Dade County is perhaps:** For a look at the situation of Miami-Dade, see (Cappucci 2019).

17 **the Netherlands figured out:** For a good overview of the Dutch perspective on climate change and sea level rise, see (Kimmelman 2017).

17 **When we realized:** When it was discovered in the 1970s that chlorofluorocarbons (CFCs)—which were in all sorts of consumer products, from hairspray to refrigerators—were damaging the ozone protecting the Earth, the response was inspiring. The international community worked together to eliminate use of ozone-depleting substances, and we began to reverse the damage. It appears now that the hole in the ozone is on its way to being fully repaired (Piper 2023).

18 **in the United States, it constituted:** This data is pulled from the Environmental Protection Agency (United States Environmental Protection Agency 2022).

19 **This number first appeared:** Nordhaus's original paper was written for the IIASA (Nordhaus 1975).

19　**was cemented in policy aspirations:** That initial 2-degree commitment was made in (European Commission 1996).

19　**One trillion tons of carbon:** CO_2 is heavier than simple carbon (makes sense, right, since it has an additional two oxygen atoms?), which is why our budget is larger if we think in terms of CO_2 rather than C. This budget was initially developed by Myles Allen and colleagues in their classic *Nature* paper (Allen, et al. 2009). It was then included in the 5th Assessment Report of the IPCC (IPCC 2014), cementing its status as a sort of countdown mechanism.

20　**the Mauna Loa Observatory:** All of this data is pulled directly from the NOAA, which you can find (along with any developments since publication) in (NOAA 2023).

21　**the IPCC put out:** The Sixth Assessment Report consists of three different documents, which are the contributions of three different Working Groups (IPCC 2021), (IPCC 2022), (IPCC 2022b).

22　**the global community needs:** According to the UNEP's most recent Emission Gap Report (United Nations Environment Programme 2022).

22　**current projections indicate:** The severity of the gap between where we're going and where we need to be was highlighted by the UN in their own news report (UN News 2022).

23　**Russia's occupying army:** For an in-depth look at the risks that were worrying scientists during the Russian occupation of the Zaporizhzhia power plant, see (Diaz-Maurin 2022).

24　**Scientists estimate that deployment:** This estimate range was provided as part of a "reality check" in a briefing paper by scientists at Imperial College London (Fajardy, et al. 2019).

24　**There are also technical issues:** A good overview of these challenges can be found in (Gough, et al. 2018).

25　**Consider, for instance, Orca:** The Orca plant was profiled by the Climate Solutions desk at *The Washington Post* in 2021, which highlighted the challenges raised in the following paragraphs (Birnbaum 2021).

26　**"The world shifts gradually":** The scenarios used by the IPCC Assessment Report were first developed by Brian O'Neill and colleagues in 2017 (O'Neill, et al. 2017) and summarized in (Riahi, et al. 2017).

27　**they estimated that humanity:** That estimate was generated as part of a larger project concerning projections of increased heat stress (Zeppetello, Raftery, and Battisti 2022).

29　**"The world follows a path":** See again (Riahi, et al. 2017).

29　**the planet is likely:** According to analyses of the pledges made at COP26 in Glasgow, universal adherence by nations to their own climate policies would likely result in 2.6–2.7C of warming by 2100. If countries went beyond adherence to their policies and met their more aggressive pledges to cut emissions by 2030, warming projections fell to 2.4C (Hausfather and Forster 2021).

30　**If we wholly fail:** Warming predictions for all of the scenarios can be found in the IPCC's 6th Assessment Report, Working Group I's Contribution (IPCC 2021, 14).

30 **"A resurgent nationalism":** Again, from Riahi and colleagues (Riahi, et al. 2017).

31 **we should expect 3.6 degrees:** This is the IPCC's best estimate of warming by 2100 under the conditions described in this scenario (IPCC 2021, 14).

31 **"finds that the international community":** For the full report see (United Nations Environment Programme 2022). The pull quote is from the landing page, which is here: https://www.unep.org/resources/emissions -gap-report-2022.

33 **resulting in ice-free Arctic summers:** There is wide variation in model predictions of when the Arctic is likely to experience its first ice-free summer. A recent paper in the journal *Climate* shows that the average of several models predicts the first ice-free Arctic summer in 2054, though the authors also show that the most likely models predict a much more pessimistic date of 2034 (Peng, et al. 2020).

33 **Scientists estimate that:** For this data, as well as predictions of heat exposure under other scenarios, see (Zeppetello, Raftery, and Battisti 2022).

33 **If we were to blow:** For this and other terrifying information about what a world of 4C warming would look like, see the report commissioned by the World Bank, *Turn Down the Heat* (World Bank 2013, 12).

34 **climate change is already responsible:** This is the finding of a recent paper in *Nature Climate Change* (Vicedo-Cabrera and Gasparrini 2021).

34 **the increased heat:** See further (Bressler 2021). Bressler further noted that, according to this projection, adding 4,434 tons of CO_2 to the atmosphere will cause one excess death by the end of the century. That means that the lifetime emissions of just four average Americans will cause the death of one person in the future. Of course, the language of "cause" here is slippery (as will be discussed in the following chapters), since the emissions by themselves do not necessarily cause any harm. They combine with trillions of other tons of emissions to lead to climate change that causes disruptions that cause harm. But the idea that such a relatively small contribution is correlated with an excess death in the future is striking.

34 **These small nations:** For a thorough overview of risks to low-lying areas from sea level rise, see the IPCC's 6th Assessment Report on risks and vulnerability (IPCC 2022).

35 **it will cause serious destruction:** As demonstrated in a study that used satellite images to estimate flood extension under warming scenarios (Tellman, et al. 2021).

35 **Recent data suggests:** See further (Kulp and Strauss 2019).

35 **which had the bad luck:** The first flood was in 2016; the second in 2018. Both were catastrophic, requiring a massive rebuilding of the entire town (Halverson 2018).

35 **the Atlantic Meridional Overturning Circulation:** For a primer on the AMOC, check out FAQ 9.3 of the IPCC's 6th Assessment Report, Working Group 1 Contribution (IPCC 2021).

36 **it will cost a cumulative:** This estimate comes from a Congressional Hearing for the Select Committee on the Climate Crisis (Liverman 2019).

36 **is greater than the sum:** As of 2021, the world's combined wealth was $431 trillion (Williams 2021).

37 **we could reduce the number:** See figure 3 of (Bressler 2021).

CHAPTER 2

41 **"The power of consumerism":** This quote is from (Monbiot 2019).

43 **a now-classic paper:** Sinnott-Armstrong's paper was originally published in (Sinnott-Armstrong 2005).

44 **When Sen. Joe Manchin:** For a look at the importance of that one vote, see (Snell 2022).

45 **Americans are some of:** These emission numbers come from the World Bank (Climate Watch 2020).

46 **"Every tonne of CO_2 emissions":** From the 6th Assessment Report (IPCC 2021, 28).

46 **The atmosphere is one layer:** Philosopher Ben Hale gives a nice overview of the complexity of the carbon cycle, and how that should affect our moral reasoning (Hale 2011).

49 **The excellent journalist:** Heglar's essay—worth reading in its entirety—is at *Vox* (Heglar 2019).

49 **In a world where:** This argument is made by Morten Fibieger Byskov (Byskov 2019), though I also hear it regularly from students and audience members when discussing this topic, so it seems to have made its way into the cultural mainstream.

50 **Climate writer David Wallace-Wells:** Wallace-Wells's full view is available in his book, *The Uninhabitable Earth* (Wallace-Wells 2019b). His particular use of this phrase can be found in his shorter piece, "Time to Panic" (Wallace-Wells 2019).

50 **"victim-blaming":** This quote is pulled from the same piece on why she doesn't care if you recycle (Heglar 2019).

50 **"Any time you are told":** Mann's book provides a sustained look at the risks of focusing on individuals to solve collective problems (Mann 2021). The quote is from an interview he did for *The Guardian* (Watts 2021).

52 **"I don't care how green":** This quote, too, is from (Heglar 2019).

CHAPTER 3

53 **"How can I begin":** This quote is from Iris Marion Young's posthumously published book, *Responsibility for Justice* (Young 2011).

55 **During that first pandemic spring:** For a flashback to that strange time, see (Ortiz 2020).

57 **I typically cannot, by myself:** This structure of certain types of actions in the modern world is characteristic of what the philosopher Judith Lichtenberg calls the "New Harms" of global collective action (Lichtenberg 2010).

58 **We raise and kill animals:** For just a glimpse at the literature supporting these claims, see Peter Singer's classic work that helped launch the animal

liberation movement (Singer 1975). Among the massive literature it spawned, see particularly the case for animal harm made in Singer and Mason's more updated work (Mason 2006), and Stuart Rachels's book chapter, "Vegetarianism" (Rachels 2011). For a more comprehensive overview of the arguments for vegetarianism, see the entry in the *Stanford Encyclopedia of Philosophy* (Doggett 2018).

58 **animal agriculture is very bad:** Evidence for the environmental benefits of a plant-based diet can be easily found. A particularly thorough and authoritative source is the EAT-Lancet Commission's report, which explores how to sustainably feed a population of 10 billion people this century (Lancet Commissions 2019).

59 **as much as 14.5 percent:** This data comes from the United Nations' Food and Agriculture Organization (Gerber, et al. 2013).

63 **Objections to vegetarian arguments:** See, for instance, discussion by Shelly Kagan (Kagan 2011) and Alistair Norcross (Norcross 2004).

63 **others objected that the pandemic:** For a particularly striking example of this reasoning applied to the Covid-19 pandemic, see German Lopez's argument in *Vox* (Lopez 2020).

CHAPTER 4

69 **"It seems to me":** This quote comes from the philosopher T. M. Scanlon's groundbreaking work, *What We Owe to Each Other* (Scanlon 1998).

69 **"a young pair of siblings":** The case of Mark and Julie comes from a study that Haidt performed with two of his colleagues on the phenomenon of "moral dumbfounding" (Haidt, Bjorklund, and Murphy 2000) but which he then goes on to report in a phenomenally influential paper describing his social intuitionist approach to moral judgment (Haidt 2001).

72 **Haidt endorses a dual-process theory:** Haidt discusses the role of the dual-process view in that original report of the study (Haidt 2001), but it also shows up prominently in his later work. In his book *The Happiness Hypothesis*, he uses the metaphor of an elephant and a rider to help explain his view of the "divided mind," with the elephant representing our emotional selves, and the rider limited in their ability to directly influence the elephant's direction (Haidt 2006).

72 **there is one group:** This point is made in Haidt's discussion of how the social-intuitionist model can helpfully explain how our reasoning and emotions fit together (Haidt 2001, 829).

73 **The dual-process theory:** Haidt admits this but seems somewhat skeptical that teaching philosophy is likely to make a large difference. He writes, "If the social intuitionist model is correct as a description of human moral judgment, it may be possible to use the model to get reasoning and intuition working more effectively together in real moral judgments. One approach would be to directly teach moral thinking and reasoning skills . . . However, attempts to directly teach thinking and reasoning in a classroom setting generally show little transfer to activities outside of the classroom (Nickerson, 1994), and because moral judgment involves 'hotter' topics than are usually dealt with in courses that attempt to teach thinking and

reasoning, the degree of transfer is likely to be even smaller" (Haidt 2001, 829). He may be right, but I've seen enough students and colleagues come to be truly changed by moral reasoning that I'm willing to bet on the effort.

CHAPTER 5

78 **"But what will become"**: From Dostoevsky's novel *The Brothers Karamazov* (Dostoevsky 1880).

79 **it is easier for:** Matthew 19:23–24. "Then Jesus said to His disciples, 'Assuredly, I say to you that it is hard for a rich man to enter the kingdom of heaven. And again I say to you, it is easier for a camel to go through the eye of a needle than for a rich man to enter the kingdom of God.'"

81 **"wouldn't it suck fallacy"**: With thanks to Justin Weinberg, circa 2006.

81 **Dmitri suggests that:** Original in Dostoevsky (Dostoevsky 1880), Sartre's paraphrase in *Existentialism Is a Humanism* (Sartre 1946).

82 **Plato wrestled with:** What follows is a summary of various parts of *The Republic*, for which I reference Alan Bloom's translation and commentary (Plato 2016).

82 **It's a running joke:** With thanks to my good friend Gordon Purves, who first told this joke to me in graduate school at the University of South Carolina (almost certainly while complaining about course reading).

83 **Glaucon started off quite suspicious:** After Socrates thrashes the bully, Thrasymachus (who argues against Socrates a version of "might makes right"), in Book I, Glaucon begins Book II by rehabilitating Thrasymachus's argument in order to push Socrates for a defense of the goodness of justice in and of itself. He says to Socrates that in the view of many, justice "seems to belong to the form of drudgery, which should be practiced for the sake of wages and the reputation that comes from opinion; but all by itself it should be fled from as something hard" (Plato 2016, Line 358a).

84 **explores the problem of moral motivation:** On Scanlon's view, the problem of moral motivation tells us much about the shape of the correct account of morality (Scanlon 1998).

85 **none of them enjoy anything:** Scanlon did offer a solution to the dilemma, positing that the reason to be moral is the reason we have to stand in a relationship of mutual justifiability with others. Although I think this is a lovely view, I've argued that it's not really a solution, since it falls much too close to the Kantian horn of the dilemma (you should be moral because being in the moral relationship is good). In my response to Scanlon, I offer my own solution based on his broader view, which is that the good of deep, meaningful relationships requires a commitment to justifiability, but I think it's fair to say that no one thinks I've definitively solved this problem either! For more details on Scanlon's view, and on my published response, see Scanlon's *What We Owe to Each Other* (Scanlon 1998) and my "How to Solve Prichard's Dilemma" (Rieder 2015).

86 **Socrates is floored:** Clearly, part of Socrates's reaction is due to the terrible norms of the time, according to which slaves don't have full moral standing, and so defending one (against one's own father!) was unthinkable.

86 **cleverly titled "Euthyphro":** "Euthyphro" is a very short dialogue, often published with other short dialogues that document the final period of Socrates's life. The version I cite here is in such a collection, translated by G.M.A. Grube (Plato 2002).

CHAPTER 6

93 **"That's just, like":** This quote is from the brilliant film *The Big Lebowski* (Coen and Coen 1998).

97 **"Tastes and colors are inarguable":** Some translators use "unquestionable" rather than "inarguable," but Sadiye's seems both more accurate and more helpful for this context.

99 **We cannot directly observe:** There is an interesting and extensive literature on whether morality is "like mathematics" in various ways, and I'm not trying to take a firm stand here. Rather, I'm just relying on what I take to be the obvious claim that math gives us an example of non-empirical facts. It is objectively true that math facts hold, even though we're not perceiving those facts with our senses (the way we do with empirical facts). And so we already accept the existence of non-empirical facts. For a deeper dive on the morality-math analogy and any implications it has for the epistemology and ontology of ethics, see Justin Clark-Doane (Clark-Doane 2014).

99 **There is much abstract philosophy:** For just a small sampling, consider the following. There is a divide between philosophers who think that moral reality is objective, mind-independent, and nonnatural (the nonnatural realists). Plato was the paradigmatic nonnatural realist (Plato 2016), but the modern revival draws inspiration from G. E. Moore (Moore 1903). There are also naturalist realists like Peter Railton, who believe that morality is real and mind-independent, but that moral facts are not spooky or different from other sorts of facts (Railton 1986). The radical opposite end of the spectrum is occupied by the moral nihilists, who deny that there are any moral facts. Most of these folks follow J. L. Mackie and identify as "error theorists," arguing that we tend to believe in moral facts but are systematically in error when we make those judgments (Mackie 1977). And in the middle are a whole slew of philosophers who think that there are moral facts, but that they must be explained in some constructed way—that is, that they are not mind-independent but exist because humans exist. This includes our old friend T. M. Scanlon, whose contractualism is a form of "moral constructivism" (Scanlon 1998).

CHAPTER 7

100 **"People do not like":** This quote comes from Keller's collection (Keller 1967).

104 **Most people asked:** According to the classic study of this phenomenon, 85 percent of people would throw the switch (Hauser, et al. 2007).

105 **you don't think you should:** In this scenario, 12 percent of participants said they would push the big man (Hauser, et al. 2007).

105 **they were originally developed:** For Foot's classic version, see (Foot 1967).

105 **some even think:** This is one of the conclusions of a pretty thorough takedown of trolley problems by Adrian Rennix and Nathan J. Robinson (Rennix and Robinson 2017).

108 *the categorical imperative:* Confusingly, Kant actually claimed that the categorical imperative could be specified in a few different ways. The "humanity as an end in itself" formulation is the one I cite in the main text because it is widely held to be the most plausible version. The core insight here can be found in many people's intuitions, as well as in many societies' policies: we are required to respect people by not treating them like things to be manipulated for our own purposes. Kant described his moral philosophy in several places. The most accessible is *Groundwork of the Metaphysics of Morals* (Kant 1785).

108 **depends on their having consented:** The importance of consent for a requirement to respect others is reflected in the entire social contract tradition. See especially John Rawls (Rawls 1971) and T. M. Scanlon (Scanlon 1998).

109 **a serious moral difference:** For a very thorough exploration of the doing/allowing distinction, see (Quinn 1981).

111 **The exact formulation:** There are many competing versions of the Doctrine of Double Effect, though they all tend to emphasize the same things. Joseph Mangan articulates the Doctrine in the following way:

> "A person may licitly perform an action that he foresees will produce a good effect and a bad effect provided that four conditions are verified at one and the same time:
>
> 1. that the action in itself from its very object be good or at least indifferent;
> 2. that the good effect and not the evil effect be intended;
> 3. that the good effect be not produced by means of the evil effect;
> 4. that there be a proportionately grave reason for permitting the evil effect" (Mangan 1949).

111 **If one does violence:** This was a reverse of Aquinas's earlier view, which was that killing in self-defense is impermissible. The double effect argument is made in the *Summa Theologica* (Aquinas 13th Century).

111 **it helps with trolleys:** Indeed, the trolley problem was developed by Phillipa Foot explicitly as a defense of the Doctrine of Double Effect (Foot 1967).

111 **what makes something a sandwich:** Shout-out to my dear friend Tony Ficarrotta, who will argue endlessly about whether a hot dog, a stuffed pita, or any other meat-and-bread combo is a sandwich! (Perhaps unsurprisingly, Tony is now a successful lawyer.)

111 **whether blowing someone up:** The example that launches this discussion concerns a group of cavers, who allow a large member of their party to lead them out of the cave, but then are trapped when the rotund fellow gets wedged in the exit. Unfortunately, water begins to rise rapidly,

threatening to drown them all if they don't dislodge the big man, and their last resort is a stick of dynamite (philosophers are exceedingly weird, I know). The example, like the trolley, was put forward by Philippa Foot (Foot 1967), and is discussed at length by John Martin Fisher, Mark Ravizza, and David Copp (Fisher, Ravizza, and Copp 1993) in their exploration of whether Warren Quinn can rescue the Doctrine of Double Effect from worries about "closeness" (Quinn 1989).

112 **a school of thought:** The *locus classicus* for virtue theory is Aristotle's *Nichomachean Ethics* (Aristotle 2014). However, after some time out of the limelight, virtue theory is enjoying something of a resurgence today, in work by scholars like Rosalind Hursthouse (Hursthouse 2002) and Christine Swanton (Swanton 2005).

112 **virtue ethics gives just:** See the extended discussion in (Hursthouse 2002).

112 **Confucius in China:** Bryan van Norden examines Confucianism as a form of virtue ethics in his book *Virtue Ethics and Consequentialism in Early Chinese Philosophy* (Norden 2007).

CHAPTER 8

115 **"The relation between moral theory":** This quote comes from a graduate-level introduction to ethical theory, written by the philosopher Mark Timmons (Timmons 2001).

115 **"Famine, Affluence, and Morality":** This article originally appeared in the inaugural volume of the journal *Philosophy and Public Affairs* (Singer 1972).

117 **the effective altruism movement:** In the early days of the effective altruism movement, moral philosophers like Singer—and those he inspired—argued for and helped to develop organizations that directed people's money to highly effective charities, where effectiveness is measured in good promoted per dollar donated. This resulted not only in the production of arguments (Singer 2015) and attempts to motivate individuals to donate their money (MacAskill 2015), but also in organizations like GiveWell and Giving What We Can. More recently, some in the effective altruism movement have changed course a bit, becoming more concerned to ensure that very many humans live for a very long time. This spinoff is called *longtermism* (MacAskill 2022) and obviously measures effectiveness quite differently from classic effective altruists. But many of those in both the traditional effective altruist camp and in the longtermist camp have been clearly influenced by Singer's utilitarian philosophy.

118 **it was developed:** Bentham laid out the classical view in (Bentham 1789).

118 **John Stuart Mill:** See especially (Mill 1861).

118 **Henry Sidgwick:** See especially (Sidgwick 1874).

118 **the answer is "nowhere":** This famous quote is pulled from Bentham's critique of all natural rights (Bentham 1795).

122 **Another famous purported counterexample:** This one comes from Judith Jarvis Thomson, in a discussion of trolley problems and related thought experiments (Thomson 1985).

123 **sacrifice all our intuitions:** Unsurprisingly, many consequentialist theories try to avoid this outcome, and so there is an extensive literature on whether modifications to the basic consequentialist framework can preserve intuitions about rights and other deontological restrictions. Perhaps the most popular is Rule Consequentialism, according to which one ought not directly promote the best consequences with every action, but should instead follow the set of rules which, if everyone followed, would lead to the best world (Hooker 2003).

124 **Singer's eventual conclusion:** Singer offers both a strong and a moderate version of his argument. On the moderate version, one must give away resources only until doing more would require sacrificing something morally significant. On the strong version, one must keep giving away resources until the point of marginal utility. Singer claims that the latter principle seems the correct one, but that he offers the moderate one to show that, even on such an "undeniable" principle of morality, a significant change to our lives is required (Singer 1972).

126 **several bioethicists engaged:** For just a sample of this long and fascinating discussion, see the contributions of Tom Beauchamp and James Childress (Beauchamp and Childress 2019), Bernie Gert (Gert 2004), John Arras (Arras 2017), and Ed Pellegrino (Pellegrino 1995).

126 **that debate continues:** Indeed, David DeGrazia and Joseph Millum published *A Theory of Bioethics* in 2021, which picks up in many ways on the debates of the previous century (DeGrazia and Millum 2021).

126 **When physicians say:** A particularly famous instance of this occurred when thirteen-year-old Jahi McMath suffered a complication after surgery for sleep apnea, resulting in cardiac arrest and an eventual declaration of brain death. Although her body was kept alive with intensive care, clinicians wanted to withdraw treatment, claiming it was futile. McMath's parents refused, however, insisting that she was still alive. They eventually won and had their daughter transferred to another hospital that would continue treating her. In 2018—five years after being declared brain dead—Jahi McMath's body finally died. The case was discussed extensively in the bioethics literature. A good example is the paper by Robert Truog (Truog 2018).

126 **how do we triage care:** Obviously, this situation arose during the Covid-19 pandemic (Emanuel, et al. 2020), though it is a classic case in public health ethics. Early in the life of ventilators, they were an expensive and rare technology, but life-saving as well. That made their allocation a key ethical question anytime need outstripped supply.

126 **intentionally infect healthy volunteers:** This form of research is called "challenge trials," and they are used regularly in animal models as well as in humans for conditions that do not pose severe risk (and for which there

is a rescue therapy). In the early stages of Covid-19, however, neither of these conditions were met. So the question was: If we could speed development of a vaccine by using a challenge model, would it be permissible to do so? Despite the fact that the typical conditions were not met, a large number of scientists and ethicists said yes (Aaronson, et al. 2020)

127 **made the era:** This is a line from the late, great bioethicist John Arras. The article in which he says this has been updated by the philosopher Jennifer Flynn, who preserves the idea (Flynn 2020). Thanks to Professor Flynn for stewarding a great piece on method in bioethics. And rest in peace, John.

127 **this methodology is called casuistry:** An early proponent of this method was Al Jonsen (Jonsen 1991).

127 **a compromise position:** It is almost impossible to overemphasize the impact that Beauchamp and Childress have had on (especially, but not exclusively, American) medical ethics. They are two of the founders of the field of modern bioethics, and the principles have been at the center of discussion for nearly half a century. In addition, many medical schools teach the principles as the basics of medical ethics to their students. Citations here are to the 8th (most recent) edition (Beauchamp and Childress 2019).

129 **part of the problem:** This criticism of utilitarianism comes from a famous essay in which Williams argues that the invasiveness of utilitarianism—were the theory true—would rob agents of their integrity (Smart and Williams 1973). We will explore this argument in more depth in later chapters.

130 **consequentialism treats all of us:** The cup analogy comes from philosopher Tom Regan, who coins it in a discussion of animal rights (Regan 1985).

131 **doesn't mean we should:** Although, it should be noted that some longtermist philosophers seem to get pretty close to endorsing this view. William MacAskill, in *What We Owe the Future*, argues that we have reasons to make people precisely because doing so adds value, and so the best world is one in which we have maximally many people doing maximally well (MacAskill 2022). In fact, he even endorses Derek Parfit's Repugnant Conclusion, according to which a very large population with people who have lives barely worth living is better—is to be morally preferred—to a much smaller one where everyone is blissfully happy (Parfit 1984). On such a view, happiness liquid is what matters, and even if each person only has a little, if there are enough people, there will be more overall happiness than if there were fewer cups, all of which are filled to the brim.

131 **Scanlon himself offers:** All of Scanlon's ideas presented in this paragraph are articulated in *What We Owe to Each Other* (Scanlon 1998).

131 **When we call someone:** Lest you think that I'm joking about "asshole" being worthy philosophical jargon, philosopher Aaron James wrote an entire book about assholes (James 2012).

132 *specifying* **the content of principles:** The process of specification was developed by Henry Richardson as a component of utilizing moral

principles (Richardson 1990) and was enthusiastically adopted by Beauchamp and Childress in subsequent editions of *Principles*. As an aside: one of the things I loved about John Arras (see the endnote above) is that he acted as a commentator on the intellectual battle about methods in bioethics. When Beauchamp and Childress got into a years-long debate with Bernie Gert, Arras noted that it was like a battle between the hedgehog and the Borg (Arras 2009). Gert was a hedgehog, putting up his armor and trying to knock down every attack. But Beauchamp and Childress—just as they did with Richardson's method of specification—took criticism and commentary and simply assimilated it. So they're the Borg from *Star Trek*—offer them commentary, and it will be assimilated. Yes, I know that was a long walk and only tangentially related, but I love that story. And besides, these are the endnotes. If you're still reading, that's kind of on you, isn't it?

CHAPTER 9

135 **"One way to confirm":** This quote comes from the paper by Sinnott-Armstrong discussed in Chapter 2 (Sinnott-Armstrong 2005).

138 **everyone who emits GHG:** For Nolt's full argument, see his essay in *Ethics, Policy & Environment* (Nolt 2011).

138 **the idea of a "threshold":** This sort of strategy is attempted by various consequentialist philosophers, including Shelly Kagan (Kagan 2011) and Alistair Norcross (Norcross 2004).

139 **Borrowing from the late philosopher:** Parfit discusses what he calls "mistakes in moral mathematics" in his groundbreaking work, *Reasons and Persons* (Parfit 1984).

139 **responsibility tracks the *risk*:** Thanks to Colin Hickey for noting that I should include risk of harm alongside actual harm.

140 **the philosophers who try:** Norcross, for instance, who has been making these arguments for decades, is known for defending a version of consequentialism called *scalar utilitarianism*. Kagan has defended act utilitarianism for many years. Nolt has argued extensively in defense of consequentialism and rational choice theory.

141 **attempts to explain the wrongness:** Nefsky explores the literature on what I've been calling the Puzzle in several publications, but explicitly makes the division I'm referring to here in (Nefsky 2015).

144 **more like being a sucker:** Interestingly, fairness generates a reason that plausibly did get off the ground at various times with respect to mitigating the harms of the Covid-19 pandemic. When mask mandates were utilized, wearing a mask was doing your fair share. When vaccines were required in order to fly or access some other goods, getting vaccinated was doing your fair share. By creating a norm and/or policy that generates widespread adoption, governments are able to generate fairness obligations.

145 **The problem with climate change:** My thinking on the topic of fairness owes much to my colleague and once-coauthor, Justin Bernstein (Rieder

and Bernstein 2020). Which is not to say he would endorse everything I say here, but his insights on duties of fairness have been very helpful as I thought through this component of the argument.

146 **an interesting and compelling literature:** In particular, see the work of Dale Jamieson, who argues that we need "green virtues" to help us determine how to live amidst climate change (Jamieson 2014).

146 **virtue ethics does tell us:** See again (Hursthouse 2002).

CHAPTER 10

149 **"There are more things":** This line is spoken by Hamlet to Horatio in Shakespeare's play *Hamlet* (Shakespeare c. 1600).

149 **the recent decision:** The judgment was handed down in June of 2022, reversing precedent set by *Roe v. Wade* and reaffirmed in *Planned Parenthood of Southeastern Pennsylvania v. Casey*, which held that the right to an abortion is protected by the U.S. Constitution (Dobbs 2022).

150 **your rights end:** This is a common understanding of Mill's famous "harm principle," which points to the fact that an act would harm others as justification for restricting one's liberty (Mill 1859).

151 **"The right to swing":** Although this is variously credited to famous thinkers like Oliver Wendell Holmes, it appears it actually originated in a law review article (Chafee 1919, 957).

151 **person is a "forensic term":** This line comes from his discussion of personal identity in *An Essay Concerning Human Understanding* (Locke 1689).

151 **It is an open:** For instance, I think it likely that some nonhuman animals have the same moral status as humans, and so are persons. This seems plausible to me in cases where an animal is so sophisticated that it likely has any property that we find morally relevant (cognitive, emotional, relational). So, some primates may be persons. Dolphins as well. Perhaps elephants. Maybe even some cephalopods. If there is alien life in the universe, it seems possible that it, too, would have personhood, even though it would not be human. On the other side, some humans lose or lack the qualities that we might think are essential for personhood. This could be true of an individual in a permanent vegetative state, for instance. And of course, the question under discussion: many people think this is true of early human life, which is biologically human but lacks most properties we think important for personhood.

152 **One prominent example:** Unsurprisingly, the view that sentience is what matters for personhood tends to be held by utilitarians. Since they think morality is about promoting happiness and minimizing suffering, what matters morally about a person coming into existence is that there is a new being that can feel. The idea that more sophisticated cognitive capacities are required for grounding moral status is often taken to be a Kantian idea, though I imagine many people who hold it have never read or thought about Kant. There are, of course, many more conceptual possibilities. For an excellent overview of different ways to ground moral status,

see Jaworska and Tannenbaum's entry in the *Stanford Encyclopedia* (Jaworska and Tannenbaum 2021).

152 **On another view:** As we will see in the discussion to come, Maggie Little advocates for a form of gradualism (Little 2008).

153 **a "future like ours":** The philosopher Don Marquis is famous for providing the "future like ours" argument for moral status, as it purports to explain the wrongness of abortion without appeal to any religious or spiritual beliefs that may not be widely shared (Marquis 1989).

154 **Thomson focuses on:** The discussion here summarizes parts of Thomson's famous article, "A Defense of Abortion" (Thomson 1971).

156 **until I met Maggie Little:** My sincere gratitude to Maggie for reading multiple drafts of this chapter and providing extensive feedback. Not everything I discuss here has made its way into her published work, and so Maggie patiently helped me to understand her philosophical worldview. Of course, if there are mistakes that nonetheless snuck their way into the chapter, those are mine and mine alone.

157 **the question of *gestation ethics*:** Maggie discusses this sort of reframing of the abortion debate in two of her seminal papers: (Little 1999) and (Little 2005).

158 **the answer is no:** This case is a variation on one given by Maggie (Little 2005). Note that in discussion of whether one is obligated to have sex or perform some other behavior, Maggie tends to say that one is "typically" not obligated. This is important because she's a moral particularist, so she doesn't endorse general moral claims. There may be some case where sex is obligatory. But it's typically not.

158 **According to what is called:** The classic citation for this idea is Wesley Hohfeld (Hohfeld 1913). Although the idea of strict correlativity has been criticized from many corners over the past hundred years, it is commonly accepted that some subset of duties is correlated with rights. If I have a right to be told the truth by you, then you have a duty to tell me the truth.

158 **members of the moral community:** The standing to demand is a core part of Darwall's contractualist theory of morality (Darwall 2009).

159 **there might be interesting cases:** Derek Parfit called these actions instances of "blameless wrongdoing" (Parfit 1984).

159 **"We do not call anything wrong":** This quote is from Mill's *Utilitarianism* (Mill 1861).

159 **there is a spectrum of "shouldiness":** My thanks to Kelly Heuer for this perfect (terrible) turn of phrase.

161 **Good sex requires:** For a great look at why consent is necessary but not sufficient for good sex, check out Quill Kukla's piece in *Aeon* (Kukla 2019).

161 **Maggie thinks that:** This idea is core to several of Maggie's papers, but is most clearly articulated in (Little 1999).

162 **such deeply intimate actions:** This quote is from (Little 2005, 153).

163 **that wouldn't be medicine:** In the United States, this point is ensconced in law. From *Schoendorff v. Society of New York Hospital*: "Every human being of adult years and sound mind has a right to determine what shall be

done with his own body; and a surgeon who performs an operation without his patient's consent commits an assault for which he is liable in damages. This is true except in cases of emergency, where the patient is unconscious and where it is necessary to operate before consent can be obtained" (*Schoendorff* 1914).

164 **There is a thorough intertwinement:** This, too, is from Maggie's paper on the duty to gestate (Little 1999). In recent years, there has been significant interesting work on the metaphysics of pregnancy, in which this broad question of the relationship between the fetus and gestating parent has been explored. The sort of "atomistic" view to which Maggie was responding is sometimes called the "containment view" of pregnancy, as it holds that the pregnant person contains an independently existing being (Smith and Brogaard 2003). Against this (probably orthodox) view, Elselijn Kingma has leaned more on the seeming fact of a "thorough intertwinement" to develop a view on which the fetus is part of the gestating parent—the "parthood view" (Kingma 2019).

164 **the moral right to decline:** This point was acknowledged in Thomson's original article (Thomson 1971).

164 **the vast majority of:** According to the CDC, more than 90 percent of all abortions take place prior to thirteen weeks of pregnancy, and 1.2 percent occur after twenty-one weeks (Jatlaoui, et al. 2019). According to the Kaiser Family Foundation, these later abortions are typically performed for one of three reasons: a serious fetal anomaly is detected; the gestating person's health is at risk; or the pregnant person discovered their pregnancy late and then was delayed as a result of policies that make later abortions more difficult to obtain (Kaiser Family Foundation 2019).

165 **Sophisticated neonatal intensive care units:** Prognoses for the smallest neonates is continuously improving. Twenty-two weeks still seems to be the earliest that infants can be born and have a chance of survival, but those born between twenty-two and twenty-seven weeks' gestational age have seen an increase in survivability in recent years (Bell, et al. 2022).

166 **the richness of:** Maggie makes this point especially clearly in the second half of "The Moral Permissibility of Abortion" (Little 2005).

166 **A reason to do:** I take this language from—who else?—T. M. Scanlon (Scanlon 1998).

167 **they are a "more basic":** Indeed, some philosophers think that reasons are the fundamental unit of normative measurement, such that all other normative concepts can be reduced to or explained in terms of reasons. See both Derek Parfit (Parfit 2011) and Mark Schroeder (Schroeder 2007) for versions of this sort of view.

170 **when I act for some reason:** This is yet another idea that I'm stealing from Scanlon—this time from his later work, *Moral Dimensions* (Scanlon 2008).

171 **some reasons can "entice":** Dancy articulates this view in a collection of essays on the philosophy of Joseph Raz (Dancy 2004), though it subsequently shows up frequently in his work.

171 **"commendatory" reasons:** Maggie articulates this view in her contribution to a collection about Dancy's philosophy (Little 2013).

171 **Juridical reasons:** This name comes from Kant's distinction between juridical and ethical duties in *The Metaphysics of Morals* (Kant 1797).

171 **philosophers sometimes call:** I believe Dancy was the first to categorize such reasons as "peremptory" in distinguishing them from "enticing" reasons (Dancy 2004).

172 **Commendatory reasons preserve moral latitude:** I take this analysis of commendatory reasons from Maggie's work with her sometimes-coauthor Coleen Macnamara (Little and Macnamara 2022).

173 **the kind and weight:** Maggie discusses gradualism in her article on the margins of personhood (Little 2008).

173 **This seems to imply:** This is appropriate, since Maggie is a moral particularist, which means she does not believe there are any universally valid moral claims. See, for instance, her early discussion of particularism in the *Hastings Center Report* (Little 2001).

CHAPTER 11

176 **"Although it may be":** This quote is from Jonathan Safran Foer's excellent book, *We Are the Weather* (Safran Foer 2020).

183 **"A mindful person":** This quote comes from Jamieson's book-length treatment of how to live amidst climate change (Jamieson 2014).

183 **Integrity is variously defined:** In the context of the Puzzle, both Marion Hourdequin (Hourdequin 2010) and Trevor Hedberg (Hedberg 2018) have argued for a duty to reduce one's carbon footprint on the basis of integrity.

184 **An interesting aspect:** This idea that there are two different ways to respond to threats like climate change comes up in Elizabeth Cripps's articulation of two different categories of resulting duties. Mimicking duties obligate us to do the thing that, were everyone to do it, would solve the problem (like reduce our emissions). Promotional duties obligate us to promote structural change, and so to work toward collective action with our individual effort. Cripps herself thinks that, despite the widespread appeal of mimicking duties, they are neither exclusive nor primary; rather, promotional duties ought to be the focus in cases like climate change (Cripps 2013).

185 **when you contribute:** See further Kutz's extended discussion of complicity in his book-length treatment of the topic (Kutz 2000).

187 **I think everyone:** For a deeper dive into what various attempted solutions to the Puzzle have gotten right and wrong, see my paper with Justin Bernstein on "contributory ethics" (Rieder and Bernstein 2020).

188 **even Sinnott-Armstrong:** This point shows up both in Sinnott-Armstrong's initial paper (Sinnott-Armstrong 2005), as well as in an updated treatment with Ewan Kingston (Kingston and Sinnott-Armstrong 2018).

191 **Jonathan Safran Foer provides:** This example is from his book *We Are the Weather* (Safran Foer 2020).

CHAPTER 12

197 **"Since it is not possible":** This quote is from Shotwell's book *Against Purity* (Shotwell 2016).

197 **The philosopher Mark Schroeder:** This example comes from Schroeder's book *Slaves of the Passions* (Schroeder 2007).

199 **attempting to seek out pleasure:** The paradox of hedonism was originally articulated by the philosopher and theologian Joseph Butler (Butler 1726).

199 **Versions of utilitarianism:** Sidgwick goes even further, suggesting that it may be better for the "vulgar" masses to reject utilitarianism. He writes, "a Utilitarian may reasonably desire, on Utilitarian principles, that some of his conclusions should be rejected by mankind generally; or even that the vulgar should keep aloof from his system as a whole, in so far as the inevitable indefiniteness and complexity of its calculations render it likely to lead to bad results in their hands" (Sidgwick 1874, 489–490).

200 **A similar move:** Peter Railton makes a version of this argument by pointing out that the rule of maximizing happiness can be either the criterion of right action or a decision procedure. If it's the criterion of right action, then an agent may well do better at acting rightly by adopting certain traits than by utilizing a maximizing decision procedure (Railton 1984).

201 **In highly industrialized societies:** Evidence for these claims was first noted back in Chapter 3. For a good overview, see Tyler Doggett's entry in the *Stanford Encyclopedia* (Doggett 2018).

203 **"An ethical approach aiming":** This quote, like the one that opens this chapter, is from *Against Purity* (Shotwell 2016).

206 **A similar case:** Williams's version can be found in the book he coauthored with Smart, *Utilitarianism: For and Against* (Smart and Williams 1973).

209 **This sort of heuristic:** This tagline comes from Pollan's famous book *In Defense of Food* (Pollan 2009).

213 **the only pressure you face:** Of course, if you believe that some reasons are merely commendatory, or enticing, then you may not face even this pressure. However, since that is not the dominant view, I won't assume it here.

214 **reminiscent of Immanuel Kant's notion:** See, for instance, Kant's discussion of the different kinds of duties in *The Metaphysics of Morals* (Kant 1797).

215 **Recalling Maggie Little's discussion:** As noted above, we need not assume all of Maggie's view to make this point. It is not crucial, for instance, to hold that there are commendatory reasons. If there are, then we get latitude automatically. But my argument here suggests that we can also generate latitude as a result of the structural features of catastrophe ethics—because we cannot respond to all reasons, so we must develop a method for choosing some to respond to.

216 **Warren Buffett, who has pledged:** Buffett pledged in 2006 to give virtually all of his wealth away, and along with Melinda and Bill Gates, has used his own promise to recruit other ultrawealthy individuals to do the

same. On the website of The Giving Pledge, he writes, "Were we to use more than 1% of my claim checks (Berkshire Hathaway stock certificates) on ourselves, neither our happiness nor our well-being would be enhanced. In contrast, that remaining 99% can have a huge effect on the health and welfare of others" (Buffett n.d.).

216 **MacKenzie Scott, who since 2019:** Scott has also signed onto The Giving Pledge (Scott n.d.).

226 **of climate activist Greta Thunberg:** Thunberg's protests on Fridays outside the Swedish Parliament eventually grew into the Fridays for Future movement and organization, which broadly tries to inspire young people to mobilize against climate inaction (Fridays for Future n.d.).

227 **"We are building a community":** Learn more about McKibben's organization at the Third Act website (Third Act n.d.).

CHAPTER 13

230 **"If children were brought":** This quote is from Schopenhauer's collection of essays (Schopenhauer 1893).

231 **If you plan to have:** As noted, this was the conclusion of Tina's dissertation (Rulli 2011). However, she also has argued for it in various places in the peer reviewed literature—for instance, in *Philosophy Compass* (Rulli 2016b).

231 **The classic case:** Indeed, the drowning-child case is classic precisely because it was used by Peter Singer in the argument discussed earlier concerning the duty to rescue people from poverty and famine (Singer 1972).

232 **Wanting your child:** Tina makes this argument in (Rulli 2016).

233 **two-parent families:** This data can be found in the 2017 report by the USDA (Lino, et al. 2017).

233 **But it could also be:** This argument is made by Stuart Rachels in *Ethical Theory and Moral Practice* (Rachels, 2014).

234 **Arthur Schopenhauer famously claimed:** See the epigraph to this chapter (Schopenhauer 1893). I'll bet Art was quite fun at parties.

234 **coming into existence is always:** This is the conclusion of his book *Better Never to Have Been* (Benatar 2006), which has spawned something of a cottage industry of philosophers responding to it.

237 **at what point each year:** All of the Global Footprint Network's data and infographics can be accessed at its website for Earth Overshoot Day (Global Footprint Network n.d.).

237 **some estimates have been:** For instance, Gretchen Daily and colleagues have argued that, until culture and technology change radically, the optimum human population is likely in the vicinity of 1.5–2 billion people (Daily, Ehrlich, and Ehrlich 1994). More recently, David Pimentel and colleagues have come to a similar estimate, arguing that a population of 2 billion people might be sustainable if we reduce consumption of the Earth's natural resources (Pimentel, et al. 2010). Theodore P. Lianos and Anastasia Pseiridis are slightly more optimistic, estimating that the earth can sustain as many 3.1 billion (Lianos and Pseiridis 2016).

238 **an incredibly invasive claim:** To be clear, even if morality rules out pro-creating, one would still have political or legal freedom to form their family how they see fit. So the issue here is not about restricting one's actual freedom of choice or movement. It's about what the rules of morality allow one to do. If you are obligated not to procreate, then you do not have the moral right to procreate, and it's that lack of moral freedom that I'm referencing here.

238 **some actions are so intimate:** I made this argument in the *Journal of Moral Philosophy* (Rieder 2014).

239 **too close to eugenic ideologies:** For anyone interested to think more about the relationship between procreative ethics and eugenics, there is a rich literature to explore. For the relationship between abortion ethics and eugenics, see Angela Davis's seminal article on the topic (Davis 1982). For the relationship between eugenics and so-called designer babies, see discussion of Julian Savalescu and John Harris's view that we are obligated to enhance our children (Sparrow 2011).

241 **Benatar thinks that the answer:** The discussion here summarizes large parts of *Better Never to Have Been* (Benatar 2006).

243 **a child would be harmed:** Jeff McMahan argues that there is a class of harms that specifically make sense of our intuition here—what he calls noncomparative harms (McMahan 2009). This feels a bit like stipulating the answer to me, but a full discussion of the issue is beyond the scope of our project here.

243 **Consider the following case:** A version of this case is used by Parfit in *Reasons and Persons* (Parfit 1984).

245 **In a slogan:** This idea has a rich history in the ethical theory debate. One of my favorite papers making this point is by Jan Narveson (Narveson 1967), who argues that if the king of a population of people called the Fervians promised to make Fervians happier, they would be rightly pissed off if he did this by annexing a very happy population, making *those people* into Fervians and thereby raising the average happiness level of all Fervi-ans. The sort of happiness increase that counts morally is an increase that is experienced by someone. More recently, William MacAskill has explicitly rejected this general point, arguing that we do, in fact, have good reason to make happy people, which is part of his argument for longtermism. This is why, as I pointed out in an earlier endnote, MacAskill embraces Parfit's Repugnant Conclusion (MacAskill 2022).

248 **scientists tried to calculate:** For the complete argument, see the paper by Paul Murtaugh and Michael Schlax (Murtaugh and Schlax 2009).

248 **One could reasonably think:** Thanks to Tina Rulli for pushing back on the Murtaugh and Schlax view and offering a couple of alternatives.

249 **The reasons that count:** I have argued for versions of this conclusion for many years now, first in the article I cite earlier concerning adoption (Rieder 2014) and then in my short book *Toward a Small Family Ethic* (Rieder 2016). A broad theme in much of my scholarship is that the strength of antinatalist arguments may not mean that their conclusion is true, but it

does mean that the arguments need to be reckoned with. There are real risks and costs to making a new person.

250 **Almost 400,000 babies:** Data is pulled from the UNICEF Data Warehouse. Annual number of births by year can be found in their data set; daily number is generated by dividing annual total by 365 (UNICEF 2023).

251 **there is a long history:** A good overview of the reasons to worry about depopulation (and policy solutions to depopulation) are covered in the UNFPA report *Depopulation* (Lutz and Gailey 2020).

251 **countries like Japan:** Indeed, although developed nations in general worry about low and falling birthrates, Japan seems to be in a state of crisis. The prime minister, Fumio Kishida, recently said that the country is "on the verge" of determining whether they could continue to function as a society with so few young people and such a large aging population (Wright 2023).

251 **This is another moral asymmetry:** This then applies to any argument that one can generate that we ought to make babies for the sake of everyone. At the extreme end, you could just believe that the continuation of the human species is a good, and so by having a child, you are doing your part. But here again, it seems morally problematic to create a whole new person for the sake of some good that's not primarily about them.

CHAPTER 14

257 **"In the absence of":** This quote is from an article that Saida Grundy wrote for *The Atlantic* (Grundy 2020).

257 **Reading groups popped up:** Kendi's (Kendi 2019) and Oluo's (Oluo 2018) books were and continue to be popular book-club choices.

258 **Johnson's concern was echoed:** Like, for instance, Saida Grundy's piece in *The Atlantic*, from which the chapter's epigraph came (Grundy 2020). Or Fabiola Cineas's piece in *Vox* (Cineas 2021), Rachel Charlene Lewis's piece at *Bitch Media* (Lewis 2020), or Claire Fallon's piece at *Huff-Post* (Fallon 2020).

259 **One form of the backlash:** My thanks to Olúfẹ́mi O. Táíwò for raising this point to me—first in discussion, and then at a talk at the Johns Hopkins Berman Institute of Bioethics titled "The Problem Is Power."

260 **Every time I choose:** My thanks to my friend and coauthor, Nabina Liebow, who alerted me to the example of choosing entertainment. She and Trip Glazer wrote about it in a paper of theirs on "White Ignorance" (Glazer and Liebow 2020).

261 **The argument here:** See the full paper in the journal *Bioethics* (Liebow and Rieder 2022).

264 **more than 100,000 people:** This tragic figure has been growing for years. When I first began working in the area of drug overdose and addiction in 2015, public health officials were already decrying an "overdose epidemic," and the number dead that year from overdose was 52,000—less than half what it was in 2021. All of this data is publicly available

through the CDC's National Vital Statistics System. Updated information can be accessed at the Vital Statistics dashboard (Ahmad, et al. 2023).

264 **We could prevent very many:** I provide the argument for this claim in my last book, *In Pain* (Rieder 2019), where I also provide extensive citations to the evidence base for each of these interventions.

266 **follow the endnotes:** A couple of books are really must-reads on this topic. For the relationship between drug use, racism, and mass incarceration, take a look at Michelle Alexander's *The New Jim Crow* (Alexander 2010). For a look at the history and a rich analysis of the harm reduction movement, see Maia Szalavitz's *Undoing Drugs* (Szalavitz 2021). For a richer understanding of addiction that helps to undermine stigma, check out Gabor Maté's *In the Realm of Hungry Ghosts* (Mate 2008). There is, of course, also an extensive academic literature on these topics. The bibliographies of all of those books, as well as my own (Rieder 2019), can point interested readers in the right direction for many topics regarding drug use, addiction, overdose, and America's current crisis.

266 **there is much to learn:** Resources are available earlier in these endnotes when we first discussed animal welfare and cruelty.

267 **Every realistic plan:** Getting exact numbers on this transition timeline is difficult, but one estimate is that the United States needs to move almost entirely to electric cars by 2050 in order to meet its climate goals (Iaconangelo 2020).

267 **"The titanic companies":** This quote is from Kara's incredible exposé of the horror show that is cobalt mining in the DRC, *Cobalt Red* (Kara 2023).

270 **What role do I:** This is similar to how the philosopher Iris Marion Young frames her response to injustice, which involves participating according to various criteria she lays out (Young 2011).

Bibliography

Aaronson, Scott, et al. 2020. "US: Challenge Trials for Covid-19." 1Day Sooner, July 15. Accessed March 17, 2023. https://www.1daysooner.org/us-open -letter.

Ahmad, F. B., J. A. Cisewski, L. M. Rossen, and P. Sutton. 2023. "Provisional Drug Overdose Death Counts." Accessed March 26, 2023. https://www .cdc.gov/nchs/nvss/vsrr/drug-overdose-data.htm.

Aiello, Leslie C. and R. I. M. Dunbar. 1993. "Neocortex Size, Group Size, and the Evolution of Language." *Current Anthropology* 34 (2): 184–193.

Alexander, Michelle. 2010. *The New Jim Crow: Mass Incarceration in the Age of Colorblindness*. New York: The New Press.

Allen, Myles R., David J. Frame, Chris Huntingford, Chris D. Jones, Jason A. Lowe, Malte Meinshausen, and Nicolai Meinshausen. 2009. "Warming Caused by Cumulative Carbon Emissions Towards the Trillionth Tonne." *Nature* 458: 1163–1166.

Aquinas, Thomas. 13th Century. *Summa Theologica*.

Aristotle. 2014. *Nichomachean Ethics*. Edited by C.D.C. Reeve. Indianapolis: Hackett.

Arras, John D. 2009. "The Hedgehog and the Borg: Common Morality in Bioethics." *Theoretical Medicine and Bioethics* 30 (1): 11–30.

Arras, John. 2017. *Methods in Bioethics: The Way We Reason Now*. Oxford: Oxford University Press.

Associated Press. 2021. "Cyprus Says August Was Hottest Month on Record Since 1983." September 3. Accessed February 2023. https://apnews.com /article/europe-environment-and-nature-climate-change-cyprus -b186baf0bb043ab6f731226cd81d2c26.

Beauchamp, Tom, and James Childress. 2019. *Principles of Biomedical Ethics*. Oxford: Oxford University Press.

Bell, Edward F., et al. 2022. "Mortality, In-Hospital Morbidity, Care Practices, and 2-Year Outcomes for Extremely Preterm Infants in the US, 2013–2018." *Journal of the American Medical Association* 327 (3): 248–263.

Benatar, David. 2006. *Better Never to Have Been: The Harm of Coming into Existence.* Oxford: Oxford University Press.

Bentham, Jeremy. 1789. *An Introduction to the Principles of Morals and Legislation.*
———. 1795. *Anarchical Fallacies; Being an Examination of the Declarations of Rights Issued During the French Revolution.*

Birnbaum, Michael. 2021. "The World's Biggest Plant to Capture CO2 from the Air Just Opened in Iceland." *Washington Post*, September 8. Accessed March 8, 2023. https://www.washingtonpost.com/climate-solutions/2021/09/08/co2-capture-plan-iceland-climeworks/.

Bressler, R. Daniel. 2021. "The Mortality Cost of Carbon." *Nature Communications* 12.

Buckley, Cara. 2022. "They Fought the Lawn. And the Lawn's Done." *New York Times*, December 14. Accessed February 2023. https://www.nytimes.com/2022/12/14/climate/native-plants-lawns-homeowners.html.

Buffett, Warren. n.d. *Warren Buffett.* Accessed March 23, 2023. https://givingpledge.org/pledger?pledgerId=177.

Butler, Joseph. 1726. *Fifteen Sermons Preached at the Rolls Chapel.* London: J. and J. Knapton.

Byskov, Morten Fibiegor. 2019. "Climate Change: Focusing on How Individuals Can Help Is Very Convenient for Corporations." *The Conversation*, January 10. Accessed March 11, 2023. https://theconversation.com/climate-change-focusing-on-how-individuals-can-help-is-very-convenient-for-corporations-108546.

Cappucci, Matthew. 2019. "Sea Level Rise Is Combining with Other Factors to Regularly Flood Miami." *Washington Post*, August 8. Accessed February 2023. https://www.washingtonpost.com/weather/2019/08/08/analysis-sea-level-rise-is-combining-with-other-factors-regularly-flood-miami/.

Chafee, Jr., Zechariah. 1919. "Freedom of Speech in War Time." *Harvard Law Review* 32 (8): 932–973.

Cineas, Fabiola. 2021. "The Lofty Goals and Short Life of the Antiracist Book Club." *Vox*, November 11. Accessed March 26, 2023. https://www.vox.com/22734080/antiracist-book-club-robin-diangelo-ibram-kendi.

Clark-Doane, Justin. 2014. "Moral Epistemology: The Mathematics Analogy." *Nous* 48 (2): 238–255.

Climate Watch. 2020. "CO2 Emissions (Metric Tons per Capita)." Accessed March 11, 2023. https://data.worldbank.org/indicator/EN.ATM.CO2E.PC?most_recent_value_desc=true.

Coen, Ethan, and Joel Coen, dirs. 1998. *The Big Lebowski.* Working Title Films.

Cripps, Elizabeth. 2013. *Climate Change and the Moral Agent: Individual Duties in an Interdependent World.* Oxford: Oxford University Press.

Daily, Gretchen, Anne Ehrlich, and Paul Ehrlich. 1994. "Optimum Human Population Size." *Population and Environment: A Journal of Interdisciplinary Studies* 15 (6).

Dancy, Jonathan. 2004. "Enticing Reasons." In *Reason and Value: Themes from the Moral Philosophy of Joseph Raz*, edited by R. Jay Wallace, Philip Pettit, Samuel Scheffler, and Michael Smith. Oxford: Oxford University Press.

Darwall, Stephen. 2009. *The Second-Person Standpoint: Morality, Respect, and Accountability.* Cambridge, MA: Harvard University Press.

Davis, Angela. 1982. "Racism, Birth Control, and Reproductive Rights." In *Women, Race and Class*, by Angela Davis, 202–271. New York: Random House.

DeGrazia, David, and Joseph Millum. 2021. *A Theory of Bioethics.* Cambridge: Cambridge University Press.

Diaz-Maurin, François. 2022. "Experts Weigh In on the Risk of Disaster at a Ukrainian Nuclear Power Plant." *The Bulletin*, August 19. Accessed February 2023. https://thebulletin.org/2022/08/experts-weigh-in-on-the-risk-of-disaster-at-a-ukrainian-nuclear-power-plant/.

Dobbs, State Health Officer of the Mississippi Department of Health, et al. v. Jackson Women's Health Organization et al. 2022. No. 19–1392 (Supreme Court of the United States, June 24).

Doggett, Tyler. 2018. "Moral Vegetarianism." *Stanford Encyclopedia of Philosophy*, September 14. Accessed March 13, 2023. https://plato.stanford.edu/entries/vegetarianism/.

Dostoevsky, Fyodor. 1880. *The Brothers Karamazov.*

Emanuel, Ezekiel J., Govind Persad, Ross Upshur, Beatriz Thome, Michael Parker, Aaron Glickman, Cathy Zhang, Connor Boyle, Maxwell Smith, and James P. Phillips. 2020. "Fair Allocation of Scarce Medical Resources in the Time of Covid-19." *New England Journal of Medicine* 382: 2049–2055.

European Commission. 1996. 1939th Council Meeting—Environment. Brussels, 25 and 26 June 1996. President: Edo Ronchi, Minister for the Environment of the Italian Republic. June 26. Accessed February 2023. https://ec.europa.eu/commission/presscorner/detail/en/PRES_96_188.

Fajardy, Mathilde, Alexandre Koberle, Niall Mac Dowell, and Andrea Fantuzzi. 2019. *BECCS Deployment: A Reality Check.* London: Imperial College London, Grantham Institute.

Fallon, Claire. 2020. "Can a Book Club Fight Racism?" *HuffPost*, August 19. Accessed March 26, 2023. https://www.huffpost.com/entry/can-a-book-club-fight-racism_n_5f3a80abc5b6e054c3fc9d44.

Fisher, John Martin, Mark Ravizza, and David Copp. 1993. "Quinn on Double Effect: The Problem of 'Closeness.'" *Ethics* 103: 707–725.

Flavelle, Christopher. 2021. "Tiny Town, Big Decisions: What Are We Willing to Pay to Fight Climate Change?" *New York Times*, March 14. Accessed February 2023. https://www.nytimes.com/2021/03/14/climate/outer-banks-tax-climate-change.html.

Flynn, Jennifer. 2020. "Theory and Bioethics." *Stanford Encyclopedia of Philosophy*, November 25. Accessed March 17, 2023. https://plato.stanford.edu/entries/theory-bioethics/.

Foot, Phillipa. 1967. "The Problem of Abortion and the Doctrine of Double Effect." *Oxford Review* 5: 5–15.

Fridays for Future. n.d. "Fridays for Future." Accessed March 23, 2023. https://fridaysforfuture.org.

Fulton, Julian, Michael Norton, and Fraser Shilling. 2019. "Water-Indexed Benefits and Impacts of California Almonds." *Ecological Indicators* 96 (1): 711–717.

Gerber, Pierre, Henning Steinfeld, Benjamin Henderson, Anne Mottet, Carolyn Opio, Jeroen Dijkman, Alessandra Falcucci, and Giuseppe Tempio. 2013. *Tackling Climate Change Through Livestock—A Global Assessment of Emissions and Mitigation Opportunities.* Rome: Food and Agriculture Organization of the United Nations (FAO).

Gert, Bernard. 2004. *Common Morality: Deciding What to Do.* Oxford: Oxford University Press.

Glazer, Trip, and Nabina Liebow. 2020. "Confronting White Ignorance: White Psychology and Rational Self-Regulation." *Journal of Social Philosophy* 52 (1): 50–71.

Global Footprint Network. n.d. "Earth Overshoot Day." Accessed March 24, 2023. https://www.overshootday.org.

-———. n.d. "How Many Earths? How Many Countries?" Accessed February 2023. https://www.overshootday.org/how-many-earths-or-countries-do-we-need/.

Gough, Clair, Samira Garcia-Freites, Christopher Jones, Sarah Mander, Brendan Moore, Cristina Pereira, Mirjam Roder, Naomi Vaughan, and Andrew Welfle. 2018. "Challenges to the Use of BECCS as a Keystone Technology in Pursuit of 1.5C." *Global Sustainability* 1 (e5): 1–9.

Greene, Joshua D. 2014. "Beyond Point-and-Shoot Morality: Why Cognitive (Neuro)Science Matters for Ethics." *Ethics* 124 (4): 695–726.

Grundy, Saida. 2020. "The False Promise of Anti-Racism Books." *Atlantic*, July 21. Accessed March 26, 2023. https://www.theatlantic.com/culture/archive/2020/07/your-anti-racism-books-are-means-not-end/614281/.

Haidt, Jonathan. 2001. "The Emotional Dog and Its Rational Tail: A Social Intuitionist Approach to Moral Judgment." *Psychological Review* 108 (4): 814–834.

-———. 2006. *The Happiness Hypothesis: Finding Modern Truth in Ancient Wisdom.* New York: Basic Books.

Haidt, Jonathan, Fredrik Bjorklund, and Scott Murphy. 2000. *Moral Dumbfounding: When Intuition Finds No Reason.* Unpublished Manuscript.

Hale, Benjamin. 2011. "Nonrenewable Resources and the Inevitability of Outcomes." *The Monist* 94 (3): 369–390.

Halverson, Jeff. 2018. "The Second 1,000-Year Rainstorm in Two Years Engulfed Ellicott City. Here's How It Happened." *Washington Post*, May 28. Accessed March 10, 2023. https://www.washingtonpost.com/news/capital-weather-gang/wp/2018/05/28/the-second-1000-year-rainstorm-in-two-years-engulfed-ellicott-city-heres-how-it-happened/.

Hauser, Mark, Fiery Cushman, Liane Young, R. Kang-Xing Jin, and John Mikhail. 2007. "A Dissociation Between Moral Judgments and Justifications." *Mind and Language* 22 (1): 1–21.

Hausfather, Zeke, and Piers Forster. 2021. *Analysis: Do COP26 Promises Keep Global Warming Below 2C?* October 11. Accessed March 8, 2023. https://

www.carbonbrief.org/analysis-do-cop26-promises-keep-global -warming-below-2c/.

Hedberg, Trevor. 2018. "Climate Change, Moral Integrity, and Obligations to Reduce Individual Greenhouse Gas Emissions." *Ethics, Policy & Environment* 21 (1): 64–80.

Heglar, Mary Annaise. 2019. "I Work in the Environmental Movement. I Don't Care If You Recycle." *Vox*, June 4. Accessed March 11, 2023. https:// www.vox.com/the-highlight/2019/5/28/18629833/climate-change-2019 -green-new-deal.

Hohfeld, Wesley N. 1913. "Some Fundamental Legal Conceptions as Applied in Judicial Reasoning." *Yale Law Journal* 23 (16).

Hooker, Brad. 2003. *Ideal Code, Real World: A Rule-Consequentialist Theory of Morality.* Oxford: Clarendon Press.

Hourdequin, Marion. 2010. "Climate, Collective Action and Individual Ethical Obligations." *Environmental Values* 19 (4): 443–464.

Hursthouse, Rosalind. 2002. *On Virtue Ethics.* Oxford: Oxford University Press.

Iaconangelo, David. 2020. "Ninety Percent of U.S. Cars Must Be Electric by 2050 to Meet Climate Goals." *Scientific American,* September 29. Accessed March 26, 2023. https://www.scientificamerican.com/article/ninety -percent-of-u-s-cars-must-be-electric-by-2050-to-meet-climate-goals/.

IPCC. 2014. *Climate Change 2014: Synthesis Report. Contribution of Working Groups I, II and III to the Fifth Assessment Report of the Intergovernmental Panel on Climate Change.* Geneva, Switzerland: IPCC.

———. 2018. *Global Warming of 1.5°C. An IPCC Special Report on the Impacts of Global Warming of 1.5°C Above Pre-Industrial Levels and Related Global Greenhouse Gas Emission Pathways, in the Context of Strengthening the Global Response to the Threat Of Climate Change.* Geneva, Switzerland: IPCC.

———. 2021. *Climate Change 2021: The Physical Science Basis. Contribution of Working Group I to the Sixth Assessment Report of the Intergovernmental Panel on Climate Change.* Cambridge, UK & New York, USA: Cambridge University Press.

———. 2022. *Climate Change 2022: Impacts, Adaptation, and Vulnerability. Contribution of Working Group II to the Sixth Assessment Report of the Intergovernmental Panel on Climate Change.* Cambridge, UK & New York, USA: Cambridge University Press.

———. 2022b. *Climate Change 2022: Mitigation of Climate Change. Contribution of Working Group III to the Sixth Assessment Report of the Intergovernmental Panel on Climate Change.* Cambridge and New York: Cambridge University Press.

James, Aaron. 2012. *Assholes: A Theory.* New York: Anchor Books.

Jamieson, Dale. 2014. *Reason in a Dark Time: Why the Struggle Against Climate Change Failed—and What It Means for Our Future.* Oxford: Oxford University Press.

Jatlaoui, Tara C., Lindsay Eckhaus, Michele G. Mandel, Antoinette Nguyen, Titilope Oduyebo, Emily Petersen, and Maura K. Whiteman. 2019.

"Abortion Surveillance—United States, 2016." *MMWR Surveillance Summary* 68: 1–41.

Jaworska, Agnieszka, and Julie Tannenbaum. 2021. "The Grounds of Moral Status." *Stanford Encyclopedia of Philosophy*, March 3. Accessed March 18, 2023. https://plato.stanford.edu/entries/grounds-moral-status/.

Johnson, Tre. 2020. "When Black People Are in Pain, White People Just Join Book Clubs." *Washington Post*, June 11. Accessed March 26, 2023. https://www.washingtonpost.com/outlook/white-antiracist-allyship-book-clubs/2020/06/11/9edcc766-abf5-11ea-94d2-d7bc43b26bf9_story.html.

Jonsen, Albert R. 1991. "Casuistry as Methodology in Clinical Ethics." *Theoretical Medicine* 12: 295–307.

Kagan, Shelly. 2011. "Do I Make a Difference?" *Philosophy & Public Affairs* 39 (2): 105–141.

Kahneman, Daniel. 2013. *Thinking Fast and Slow.* New York: Farrar, Straus and Giroux.

Kaiser Family Foundation. 2019. "Abortions Later in Pregnancy." December 5. Accessed March 21, 2023. https://www.kff.org/womens-health-policy/fact-sheet/abortions-later-in-pregnancy.

Kant, Immanuel. 1785. *Groundwork of the Metaphysics of Morals.*
———. 1797. *The Metaphysics of Morals.*

Kara, Siddarth. 2023. *Cobalt Red: How the Blood of Congo Powers Our Lives.* New York: St. Martin's Press.

Keller, Helen. 1967. *Helen Keller: Her Socialist Years.* New York: International Publishers.

Kendi, Ibram X. 2019. *How to Be an Antiracist.* New York: One World.

Kimmelman, Michael. 2017. "The Dutch Have Solutions to Rising Seas. The World Is Watching." *New York Times,* June 15. Accessed March 16, 2023. https://www.nytimes.com/interactive/2017/06/15/world/europe/climate-change-rotterdam.html.

King, Stephen. 1978. *The Stand.* New York: Doubleday.

Kingma, Elselijn. 2019. "Were You a Part of Your Mother?" *Mind* 128 (511): 609–646.

Kingston, Ewan, and Walter Sinnott-Armstrong. 2018. "What's Wrong with Joyguzzling?" *Ethical Theory and Moral Practice* 21 (1): 169–186.

Kukla, Quill. 2019. "Sex Talk." Aeon, February 4. Accessed March 20, 2023. https://aeon.co/essays/consent-and-refusal-are-not-the-only-talking-points-in-sex.

Kulp, Scott A., and Benjamin H. Strauss. 2019. "New Elevation Data Triple Estimates of Global Vulnerability to Sea-Level Rise and Coastal Flooding." *Nature Communications* 10.

Kutz, Christopher. 2000. *Complicity: Ethics and Law for a Collective Age.* Cambridge: Cambridge University Press.

Lancet Commissions. 2019. "Food in the Anthropocene: The EAT-Lancet Commission on Healthy Diets from Sustainable Food Systems." *The Lancet* 393 (10170): 447–492.

Lewis, Rachel Charlene. 2020. "Anti-Racist Readers Are Plaguing the DMs of Black People." Bitch Media, June 26. Accessed March 26, 2023. https://

www.bitchmedia.org/article/very-online/anti-racism-book-clubs-put
-more-work-on-black-people.

Lianos, Theodore P., and Anastasia Pseiridis. 2016. "Sustainable Welfare and
Optimum Population Size." *Environment, Development, and Sustainability* 18: 1679–1699.

Lichtenberg, Judith. 2010. "Negative Duties, Positive Duties, and the 'New
Harms.'" *Ethics* 120 (3): 557–578.

Liebow, Nabina K., and Travis N. Rieder. 2022. "'What Can I Possibly Do?':
White Individual Responsibility for Addressing Racism as a Public
Health Crisis." *Bioethics* 36 (3): 274–282.

Lino, Mark, Kevin Kuczynski, Nestor Rodriguez, and TusaRebecca Schap.
2017. *Expenditures on Children by Families, 2015.* Miscellaneous Report
No. 1528-2015, Washington, D.C.: U.S. Department of Agriculture,
Center for Nutrition Policy and Promotion.

Little, Margaret O. 1999. "Abortion, Intimacy, and the Duty to Gestate." *Ethical
Theory and Moral Practice* 2 (3): 295–312.

———. 2005. "The Moral Permissibility of Abortion." In *Contemporary Debates in Applied Ethics,* edited by Andrew I. Cohen and Christopher Wellman, 27–40. Blackwell.

———. 2008. "Abortion and the Margins of Personhood." *Rutgers Law Journal*
39: 331–348.

Little, Margaret Olivia. 2001. "On Knowing the 'Why': Particularism and
Moral Theory." *Hastings Center Report* 31 (4): 32–40.

———. 2013. "In Defence of Non-Deontic Reasons." In *Thinking About Reasons: Themes from the Philosophy of Jonathan Dancy,* edited by David
Bakhurst, Margaret Olivia Little, and Brad Hooker, 112–136. Oxford:
Oxford University Press.

Little, Margaret Olivia, and Coleen Macnamara. 2022. "The Latitude-
Preserving Nature of Commendatory Reasons." In *Oxford Studies in Metaethics, Volume 17,* edited by Russ Shafer-Landau, 54–76. Oxford: Oxford
University Press.

Liverman, Diana. 2019. *Hearing on April 30, 2019: "Solving the Climate Crisis:
Drawing Down Carbon and Building Up the American Economy."* Washington, D.C.: U.S. Government Publishing Office. Accessed March 10, 2023.
https://www.congress.gov/116/meeting/house/109329/documents
/HHRG-116-CN00-20190430-QFR004.pdf.

Locke, John. 1689. *An Essay Concerning Human Understanding.*

Lopez, German. 2020. "Why America Can't Rely Solely on Individuals to Stop
Covid-19," *Vox,* December 21. Accessed March 13, 2023. https://www
.vox.com/future-perfect/22179222/covid-19-coronavirus-pandemic
-individualism-collectivism.

Lutz, Wolfgang, and Nicholas Gailey. 2020. *Depopulation as a Policy Challenge
in the Context of Global Demographic Trends.* UNDP Serbia.

MacAskill, William. 2015. *Doing Good Better: How Effective Altruism Can Help
You Make a Difference.* New York: Avery.

———. 2022. *What We Owe the Future.* New York: Basic Books.

Mackie, J. L. 1977. *Ethics: Inventing Right and Wrong.* Harmondsworth: Penguin.

Mangan, Joseph. 1949. "An Historical Analysis of the Principle of Double Effect." *Theological Studies* 10: 41–61.

Mann, Michael E. 2021. *The New Climate War: The Fight to Take Back Our Planet*. New York: PublicAffairs.

Marinova, Dora, and Diana Bogueva. 2020. "Which 'Milk' Is Best for the Environment? We Compared Dairy, Nut, Soy, Hemp and Grain Milks." *The Conversation*, October 13. Accessed February 2023. https://theconversa tion.com/which-milk-is-best-for-the-environment-we-compared-dairy -nut-soy-hemp-and-grain-milks-147660.

Marquis, Don. 1989. "Why Abortion Is Immoral." *The Journal of Philosophy* 86 (4): 183–202.

Mason-D'Croz, Daniel, Anne Barnhill, Justin Bernstein, Jessica Bogard, Gabriel Dennis, Peter Dixon, Jessica Fanzo, et al. 2022. "Ethical and Economic Implications of the Adoption of Novel Plant-Based Beef Substitutes in the USA: A General Equilibrium Modelling Study." *The Lancet: Planetary Health* 6 (8): E658—E669.

Maté, Gabor. 2008. *In the Realm of Hungry Ghosts: Close Encounters with Addiction*. Berkeley, CA: North Atlantic Books.

McMahan, Jeff. 2009. "Asymmetries in the Morality of Causing People to Exist." In *Harming Future Persons*, edited by David Wasserman and Melinda Roberts, 49–68. Dordrecht: Springer.

Mill, John Stuart. 1859. *On Liberty.*

———. 1861. *Utilitarianism.*

Monbiot, George. 2019. "The Big Polluters' Masterstroke Was to Blame the Climate Crisis on You and Me." *Guardian,* October 9. Accessed March 13, 2023. https://www.theguardian.com/commentisfree/2019/oct/09/pol luters-climate-crisis-fossil-fuel.

Moore, G. E. 1903. *Principia Ethica*. Cambridge: Cambridge University Press.

Murtaugh, Paul A., and Michael G. Schlax. 2009. "Reproduction and the Carbon Legacies of Individuals." *Global Environmental Change* 19: 14–20.

Narveson, Jan. 1967. "Utilitarianism and New Generations." *Mind* 76 (301): 62–72.

Nefsky, Julia. 2015. "Fairness, Participation, and the Real Problem of Collective Harm." Vol. 5, in *Oxford Studies in Normative Ethics*, edited by Mark Timmons, 245–271. Oxford: Oxford University Press.

NOAA. 2023. "Trends in Atmospheric Carbon Dioxide." February 6. Accessed February 2023. https://gml.noaa.gov/ccgg/trends/mlo.html.

Nolt, John. 2011. "How Harmful Are the Average American's Greenhouse Gas Emissions?" *Ethics, Policy & the Environment* 14 (1): 3–10.

Norcross, Alastair. 2004. "Puppies, Pigs, and People: Eating Meat and Marginal Cases." *Philosophical Perspectives* 18: 229–245.

Norden, Bryan van. 2007. *Virtue Ethics and Consequentialism in Early Chinese Philosophy*. Cambridge: Cambridge University Press.

Nordhaus, William D. 1975. *Can We Control Carbon Dioxide? IIASA Working Paper*. Laxenburg, Austria: IIASA.

Oluo, Ijeoma. 2018. *So You Want to Talk About Race?* New York: Seal Press.

O'Neill, Brian C., Elmar Kriegler, Kristie L. Ebi, Eric Kemp-Benedict, Keywan Riahi, Dale S. Rothman, Bas J. van Ruijven, et al. 2017. "The Roads Ahead: Narratives for Shared Socioeconomic Pathways Describing World Futures in the 21st Century." *Global Environmental Change* 42: 169–180.

Ortiz, Aimee. 2020. "Man Who Said, 'If I Get Corona, I Get Corona,' Apologizes." *New York Times*, March 24. Accessed March 13, 2023. https://www.nytimes.com/2020/03/24/us/coronavirus-brady-sluder-spring-break.html.

Parfit, Derek.1984. *Reasons and Persons*. Oxford: Oxford University Press.

———. 2011. *On What Matters (Vols. 1 and 2)*. Oxford: Oxford University Press.

Pattison, Pete, and Niamh McIntyre. 2021. "Revealed: 6,500 Migrant Workers Have Died in Qatar Since World Cup Awarded." *Guardian,* February 23. Accessed February 2023. https://www.theguardian.com/global-development/2021/feb/23/revealed-migrant-worker-deaths-qatar-fifa-world-cup-2022.

Pellegrino, Ed. 1995. "Toward a Virtue-Based Normative Ethics for the Health Professions." *Kennedy Institute of Ethics Journal* 5 (3): 253–277.

Peng, Ge, Jessica L. Matthews, Muyin Wang, Russell Vose, and Liqiang Sun. 2020. "What Do Global Climate Models Tell Us About Future Arctic Sea Ice Coverage Changes?" *Climate* 8 (1).

Pimentel, David, Michele Whitecraft, Zachary R. Scott, Leixin Zhao, Patricia Satkiewicz, Timothy J. Scott, Jennifer Phillips, et al. 2010. "Will Limited Land, Water, and Energy Control Human Population Numbers in the Future?" *Human Ecology* 38 (5): 599–611.

Piper, Kelsey. 2023. "Why the Ozone Hole Is on Track to Be Healed by Mid-Century." *Vox,* January 10. Accessed March 16, 2023. https://www.vox.com/future-perfect/22686105/future-of-life-ozone-hole-environmental-crisis-united-nations-cfcs.

Plato. 2002. "Euthyphro." In *Five Dialogues: Euthyphro, Apology, Crito, Meno, Phaedo*. Indianapolis: Hackett.

———. 2016. *The Republic of Plato*. Edited by Allan Bloom. New York: Basic Books.

Pollan, Michael. 2009. *In Defense of Food: An Eater's Manifesto*. New York: Penguin.

Poore, Joseph, and Thomas Nemecek. 2018. "Reducing Food's Environmental Impacts Through Producers and Consumers." *Science* 360 (6392): 987–992.

Quinn, Warren S. 1981. "Actions, Intentions, and Consequences: The Doctrine of Doing and Allowing." *Philosophical Review* 98 (3): 287–312.

———. 1989. "Actions, Intentions, and Consequences: The Doctrine of Double Effect." *Philosophy and Public Affairs* 18: 334–351.

Rachels, Stuart. 2011. "Vegetarianism." In *The Oxford Handbook of Animal Ethics*, edited by Tom Beauchamp and R. G. Frey, 877–904. New York: Oxford University Press.

———. 2014. "The Immorality of Having Children." *Ethical Theory and Moral Practice* 17 (3): 567–582.

Railton, Peter. 1984. "Alienation, Consequentialism, and the Demands of Morality." *Philosophy & Public Affairs* 13 (2): 134–171.

———. 1986. "Moral Realism." *The Philosophical Review* 95 (2): 163–207.

Rawls, John. 1971. *A Theory of Justice*. Cambridge, MA: Harvard University Press.

Regan, Tom. 1985. "The Case for Animal Rights." In *In Defense of Animals*, edited by Peter Singer, 13–26. New York: Basil Blackwell.

Rennix, Adrian, and Nathan J. Robinson. 2017. "The Trolley Problem Will Tell You Nothing Useful About Morality." *Current Affairs*, November 3. Accessed November 3, 2023. https://www.currentaffairs.org/2017/11/the-trolley-problem-will-tell-you-nothing-useful-about-morality.

Riahi, Keywan, et al., 2017. "The Shared Socioeconomic Pathways and Their Energy, Land Use, and Greenhouse Gas Emissions Implications: An Overview." *Global Environmental Change* 42: 153–168.

Richardson, Henry. 1990. "Specifying Norms as a Way to Resolve Concrete Ethical Problems." *Philosophy and Public Affairs* 19 (4): 279–310.

Rieder, Travis N. 2014. "Procreation, Adoption, and the Contours of Obligation." *Journal of Applied Philosophy* 32 (3): 293–309.

———. 2015. "How to Solve Prichard's Dilemma: A Complex Contractualist Account of Moral Motivation." *Journal of Ethics & Social Philosophy* 9 (1).

———. 2016. *Toward a Small Family Ethic: How Overpopulation and Climate Change Are Affecting the Morality of Procreation*. Springer.

———. 2019. *In Pain: A Bioethicist's Personal Struggle with Opioids*. New York: HarperCollins.

Rieder, Travis, and Justin Bernstein. 2020. "The Case for 'Contributory Ethics': Or How to Think About Individual Morality in a Time of Global Problems." *Ethics, Policy & Environment* (3): 299–319.

Roser, Max, Hannah Ritchie, Esteban Ortiz-Ospina, and Lucas Rodés-Guirao. 2013. *World Population Growth*. Accessed February 2023. https://ourworldindata.org/world-population-growth.

Rosoff, Matt. 2020. "California's Raging Wildfires: Bay Area Skies Turn an Eerie Orange." CNBC, September 9. Accessed February 2023. https://www.cnbc.com/2020/09/09/california-wildfires-turn-bay-area-skies-an-eerie-orange.html.

Rulli, Tina. 2011. "The Duty to Adopt." PhD dissertation, Yale University.

———. 2016. "Preferring a Genetically-Related Child." *Journal of Moral Philosophy* 13 (6): 669–698.

———. 2016b. "The Ethics of Procreation and Adoption." *Philosophy Compass* 11 (6): 305–315.

Safran Foer, Jonathan. 2020. *We Are the Weather: Saving the Planet Begins at Breakfast*. New York: Picador.

Santino, Andrew, dir. *Andrew Santino: Cheeseburger*. 2023. Netflix.

Sartre, Jean-Paul. 1946. *Existentialism Is a Humanism*.

Scanlon, T. M. 1998. *What We Owe to Each Other*. Cambridge, MA: Harvard University Press.

———. 2008. *Moral Dimensions: Permissibility, Meaning, Blame*. Cambridge, MA: Harvard Universty Press.

Schoendorff v. Society of New York Hospital. 1914. 105 N.E. 92 (New York Court of Appeals, April 14).

Schopenhauer, Arthur. 1893. *Studies in Pessimism: A Series of Essays.*

Schroeder, Mark. 2007. *Slaves of the Passions.* Oxford: Oxford University Press.

Scott, MacKenzie. n.d. "MacKenzie Scott." Giving Pledge. Accessed March 23, 2023. https://givingpledge.org/pledger?pledgerId=393.

Shakespeare, William. c. 1600. *The Tragedy of Hamlet, Prince of Denmark.*

Shotwell, Alexis. 2016. *Against Purity: Living Ethically in Compromised Times.* Minneapolis: University of Minnesota Press.

Sidgwick, Henry. 1874. *The Methods of Ethics.*

Singer, Peter. 1972. "Famine, Affluence, and Morality." *Philosophy and Public Affairs* 1 (3): 229–243.

———. 1975. *Animal Liberation.* New York: Harper Collins.

———. 2015. *The Most Good You Can Do: How Effective Altruism Is Changing Ideas About Living Ethically.* New Haven: Yale University Press.

Singer, Peter, and Jim Mason. 2006. *The Ethics of What We Eat.* Emmaus, PA: Rodale.

Sinnott-Armstrong, Walter. 2005. "It's Not My Fault: Global Warming and Individual Moral Obligation." In *Perspectives on Climate Change: Science, Economics, Politics, Ethics,* edited by Walter Sinnott-Armstrong and Richard Howarth, 293–315. Elsevier.

Smart, J.J.C., and Bernard Williams. 1973. *Utilitarianism: For and Against.* Cambridge: Cambridge University Press.

Smith, Barry, and Berit Brogaard. 2003. "Sixteen Days." *Journal of Medicine and Philosophy* 28: 45–78.

Snell, Kelsey. 2022. "After Spiking Earlier Talks, Manchin Agrees to a New Deal on Climate and Taxes." NPR, July 27. Accessed March 11, 2023. https://www.npr.org/2022/07/27/1114108340/manchin-deal-inflation-reduction-act.

Sparrow, Robert. 2011. "A Not-So-New Eugenics: Harris and Savalescu on Human Enhancement." *Hastings Center Report* 41 (1): 32–42.

Swanton, Christine. 2005. *Virtue Ethics: A Pluralistic View.* Oxford: Oxford University Press.

Szalavitz, Maia. 2021. *Undoing Drugs: How Harm Reduction Is Changing the Future of Drugs and Addiction.* New York: Hachette.

Tellman, B., J. A. Sullivan, C. Kuhn, A. J. Kettner, C. S. Doyle, G. R. Brakenridge, T. A. Erickson, and D. A. Slayback. 2021. "Satellite Imaging Reveals Increased Proportion of Population Exposed to Floods." *Nature* 596: 80–86.

Third Act. n.d. "Third Act." Accessed March 23, 2023. https://thirdact.org.

Thomson, Judith Jarvis. 1971. "A Defense of Abortion." *Philosophy and Public Affairs* 1 (1): 47–66.

———. 1985. "The Trolley Problem." *Yale Law Journal* 94 (6): 1395–1415.

Timmons, Mark. 2001. *Moral Theory: An Introduction.* New York: Rowman & Littlefield.

Truog, Robert D. 2018. "Lessons from the Case of Jahi McMath." *Hastings Center Report* 48 (6): S70—S73.

UN News. 2022. "Countries' Climate Promises Still Not Enough to Avoid Catastrophic Global Warming: UN Report." United Nations, October 26. Accessed February 2023. https://news.un.org/en/story/2022/10/112 9892.

UNICEF. 2023. "UNICEF Data Warehouse." Accessed March 26, 2023. https://data.unicef.org/resources/data_explorer/unicef_f/?ag=UNI CEF&df=GLOBAL_DATAFLOW&ver=1.0&dq=WORLD.DM _BRTS..&startPeriod=2013&endPeriod=2023.

United Nations Environment Programme. 2022. *Emissions Gap Report 2022.* UNEP, October 27. Accessed March 8, 2023. https://www.unep.org/re sources/emissions-gap-report-2022.

United States Environmental Protection Agency. 2022. *Overview of Greenhouse Gases.* May 16. Accessed February 2023. EPA, https://www.epa.gov /ghgemissions/overview-greenhouse-gases.

Vicedo-Cabrera, A. M., and A. Gasparrini. 2021. "The Burden of Heat-Related Mortality Attributable to Recent Human-Induced Climate Change." *Nature Climate Change* 11: 492–500.

Wallace-Wells, David. 2019. "Time to Panic." *New York Times*, February 16. Accessed March 11, 2023. https://www.nytimes.com/2019/02/16/opinion /sunday/fear-panic-climate-change-warming.html.

———. 2019b. *The Uninhabitable Earth: Life After Warming.* New York: Tim Duggan Books.

Watts, Jonathan. 2021. "Climatologist Michael E Mann: 'Good People Fall Victim to Doomism. I Do Too Sometimes.'" *Guardian*, February 27. Accessed March 11 2023. https://www.theguardian.com/environment /2021/feb/27/climatologist-michael-e-mann-doomism-climate-crisis -interview.

Williams, Ollie A. 2021. "World's Wealth Hits Half a Quadrillion Dollars." *Forbes*, June 10. https://www.forbes.com/sites/oliverwilliams1/2021/06 /10/worlds-wealth-hits-half-a-quadrillion-dollars/?sh=9064525309dc.

World Bank. 2013. *Turn Down the Heat: Climate Extremes, Regional Impacts, and the Case for Resilience. A Report for the World Bank by the Potsdam Institute for Climate Impact Research and Climate Analytics.* Washington, D.C.: World Bank.

Wright, George. 2023. "Japan PM Says Country on Brink over Falling Birth Rate." BBC, January 23. Accessed March 26, 2023. https://www.bbc .com/news/world-asia-64373950.

Young, Iris Marion. 2011. *Responsibility for Justice.* Oxford: Oxford University Press.

Zeppetello, Lucas R. Vargas, Adrian E. Raftery, and David S. Battisti. 2022. "Probabilistic Projections of Increased Heat Stress Driven by Climate Change." *Communications Earth & Environment* 3 (183).

Index

INDEX

About the Author

Travis Rieder, PhD, is an associate research professor at the Johns Hopkins Berman Institute of Bioethics, where he directs the Master of Bioethics degree program. He holds secondary appointments in the Departments of Philosophy and Health Policy & Management, as well as the Center for Public Health Advocacy. His first book, a memoir of opioid dependence and withdrawal, was named an NPR Best Book of 2019, and his TED Talk on the same topic has been viewed more than 2.5 million times. He has been interviewed by Terry Gross on *Fresh Air*, and his opinion writing has appeared in *The New York Times*, *The Wall Street Journal*, and *USA Today*.